Study Guide

Roger LeRoy Miller

David VanHoose

to accompany

ECONOMICS TODAY

The Macro View

Twelfth Edition

Study Guide
to accompany

ECONOMICS TODAY
The Macro View
Twelfth Edition

Roger LeRoy Miller
Institute for University Studies,
Arlington, Texas

David VanHoose
Baylor University

PEARSON
Addison
Wesley

Boston San Francisco New York
London Toronto Sydney Tokyo Singapore Madrid
Mexico City Munich Paris Cape Town Hong Kong Montreal

Reproduced by Pearson Addison-Wesley from Microsoft® Word files supplied by the author.

Copyright © 2004 Pearson Education, Inc.
Publishing as Pearson Addison-Wesley, 75 Arlington Street, Boston MA 02116

All rights reserved. Printed in the United States of America.

ISBN   0-321-19423-3

3 4 5 6 CRS 06 05 04

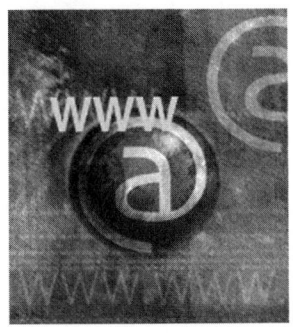

TABLE OF CONTENTS

Preface: To the Student v
Chapter 1: The Nature of Economics 1
Chapter 2: Scarcity and the World of Trade-offs 8
Chapter 3: Demand and Supply 19
Chapter 4: Extensions of Demand and Supply Analysis 33
Chapter 5: The Public Sector and Public Choice 43
Chapter 6: Taxes, Transfers, and Public Spending 59
Chapter 7: The Macroeconomy: Unemployment, Inflation, and Deflation 71
Chapter 8: Measuring the Economy's Performance 84
Chapter 9: Global Economic Growth and Development 96
Chapter 10: Real GDP and the Price Level in the Long Run 103
Chapter 11: Classical and Keynesian Macro Analysis 112
Chapter 12: Consumption, Income, and the Multiplier 125
Chapter 13: Fiscal Policy 141
Chapter 14: Money, Banking, and Central Banking 151
Chapter 15: Money Creation and Deposit Insurance 162
Chapter 16: Domestic and International Dimensions of Monetary Policy 176
Chapter 17: Stabilization in an Integrated World Economy 190
Chapter 18: Policies and Prospects for Global Economic Growth 203
Chapter 32: Comparative Advantage and the Open Economy 213
Chapter 33: Exchange Rates and the Balance of Payments 222

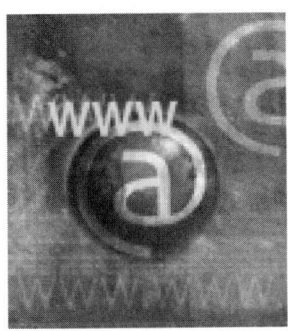

PREFACE TO THE STUDENT

This *Study Guide* is designed to help you read and understand *Economics Today*, 12th Edition. Lesson one in economics is that nothing is free; a price must be paid for every activity that is performed and for every good or service that is obtained. The price that you must pay to learn economics certainly includes the course tuition and the costs of your textbook and this *Study Guide*. The price for learning economics also includes the time and effort that it will take you to read and think about this discipline. But the benefits are, potentially, enormous.

If you really learn economics you will achieve your immediate objective, which is (presumably) to make a good grade. More importantly—for the long run—if you learn to think like an economist, you will gain some very crucial insights into human behavior. You should never forget that economics is, first and foremost, a study of human behavior. An understanding of human behavior is a necessary prerequisite to attaining your personal goals and to fulfilling any goals that you might have of helping other people. Good intentions are not enough; if you really want to help people you must understand how they are likely to respond to specific policies. And economics has proven to be an invaluable aid in understanding human motivation.

HOW THIS *STUDY GUIDE* CAN HELP YOU

This student guide can help you to maximize your learning, subject to constraints on the amount of time that you can allot to this course. There are at least five specific ways in which you can benefit from this guide.

1. The *Study Guide* can help you decide what topics are the most important. Because there are so many topics analyzed in each chapter (in *all* principles of economics textbooks) many students become confused about what is essential and what is not. You can't learn everything; the *Study Guide* can help you concentrate on the crucial topics in each chapter.

2. If you are forced to miss a class or two (we strongly recommend that you attend class regularly, but we realize that emergencies do arise), you can use this *Study Guide* to help you learn the material discussed in your absence.

3. There is a good chance that the questions you are required to answer in this *Study Guide* are representative of the types of questions that you will be asked during examinations.

4. You can use this *Study Guide* to help you review for exams.

5. Finally, this *Study Guide* can help you decide whether you really do understand the material. Don't wait until exam time to find out!

Ultimately, the way to learn economics is by reading your textbook and *thinking about the theories*. You should try to express the analysis in your own words and to apply the theory to real world circumstances. This *Study Guide* can't teach you to think like an economist—you'll have to learn to do that on your own. It can, however, provide feedback on your progress; if you can answer the questions and solve the problems, then you will know that you are on the right path.

THE CONTENTS OF THE *STUDY GUIDE*

Economics is considered to be a rather difficult subject because (a) it is theoretical in nature, (b) it uses a specialized jargon, or vocabulary, and (c) it takes (most people) much time and effort to learn. We who are economists, however, believe that our discipline is exciting and provides tremendous insights into human behavior. Your text and this *Study Guide* have been written for the precise purpose of helping you to learn economics. We always try to keep you, the student, in mind.

Before we indicate to you how we think you can best utilize this *Study Guide* to achieve your goals we want to indicate to you what it contains. Most chapters include the following sections.

1. Learning Objectives

Here we list approximately ten things that you should be able to do after you have completed the chapter.

2. Chapter Outline

This section presents a sentence outline for the chapter; it provides you with a quick overview of the contents of the chapter and it includes only the most important topics.

3. Key Terms

This section provides a list of the most important terms used in the text chapter; these terms are crucial to your understanding and each is defined in the glossary at the end of its *Study Guide* chapter.

4. Key Concepts

This is a list of the most important theoretical concepts used in the chapter; these too are explained in the end-of-chapter glossary.

5. Completion Questions

This set of short answer "fill-in-the-blank" questions is intended to test your knowledge of key terms, key concepts, and facts. Some will require an application of the theoretical concepts contained in the text.

6. True-False Questions

This is another objective test to help you see if you understand the main issues in the chapter. We also explain what is wrong with each false statement. We believe that this will be very helpful to you.

7. Multiple-Choice Questions

The numerous multiple-choice questions in each chapter are another objective test to help you decide whether or not you need to spend more time and effort on the chapter at hand.

8. Working with Graphs

Because graphs are so central to the study of economics, we decided to allocate an entire section (in those chapters where applicable) devoted to helping you interpret graphs. We believe that if you can master graphical analysis, the rest of economics will follow easily. The questions appearing in this section are the actual questions that we pose to our students in the classroom; teachers know that if students cannot answer such questions there is no point in moving on.

9. Problems

This section requires you to take a pencil in hand and solve specific problems that depend on your knowledge of the chapter's contents. Often students believe that they understand certain concepts, but then they cannot solve related problems. By working on these questions you will see the concepts in another light, and a deeper understanding will emerge.

10. Answers

We have placed the answers to the *Study Guide* at the end of each chapter, and not at the end of the book. We believe that you will find this convenient.

11. Chapter Glossary

Because economics vocabulary is so important, we decided to add it to the student guide. And we have placed it where it is the most useful: at the end of the relevant chapter, not at the end of the book.

HOW TO USE THIS STUDY GUIDE

What follows is a recommended strategy for improving your grade. It may seem like an awful lot of work, but the payoffs will be high. Try the whole program for the first three or four chapters. If you feel you can skip some steps safely, then try doing so and see what happens. After all, only you can know your capabilities and individual circumstances. We do urge you, however, to give this approach a chance.

For each chapter we recommend that you follow the sequence of steps below.

1. Read the introduction, the learning objectives, the sentence outline, and the lists of key terms and concepts in this *Study Guide*; follow any study suggestions offered in the introduction.

2. Read the Chapter Summary in your text at the end of the chapter.

3. Read about half the textbook chapter (unless it is very long), being sure to underline only the most important points which you should be able to recognize after having read two chapter outlines. Put a check mark by that material that you don't understand.

4. If you find the textbook chapter easy to understand, you might want to finish reading it. Otherwise, rest for a sufficient period (you can be the judge of how long it takes you to be refreshed) before you read the second half of the chapter. Again be sure to underline only the most important points and to put a checkmark by the material you find difficult to understand.

5. After you have completed the entire textbook chapter, take a break. Then read only what *you* have underlined, throughout the entire chapter.

6. Now concentrate on the difficult material, by which you have left checkmarks. Reread this material and *think about it*; you will find that it is very exciting to figure out difficult material on your own.

7. Read each of the chapter preview questions in the textbook and write out your answer. After you have finished, compare your answers to the answers to those questions provided by your author at the end of the chapter.

8. Find the comparable chapter in this *Study Guide* and answer the completion questions, the true-false questions, the multiple-choice questions, the problems, and the working with graphs questions. Compare your answers with the answers provided at the back of the *Study Guide* chapter. Make a note of the questions you have missed and find the page(s) in your textbook upon which these *Study Guide* questions are based. If you still don't understand—ask your teacher or your student teaching assistant. (If you decided that our answers are wrong, then by all means write and tell us.)

9. Re-read the learning objectives in this *Study Guide* and decide if you are able to achieve each of these objectives.

10. Before your examination, study your class notes. Reread the *Study Guide* outline, then re-do the completion questions, the true-false questions, the multiple-choice questions, and the problems. Compare your answer with the answer at the back of the appropriate chapter in this guide. Identify your problem areas and re-read the relevant pages in the book. Think through the answers on your own. If you still can't understand the analysis, ask your teacher or your student teaching assistant for help. (Be sure to let your teacher know that you have tried to answer the questions on your own.)

If you have followed the strategy outlined above, you should feel sufficiently confident and be relaxed to do well on your exam.

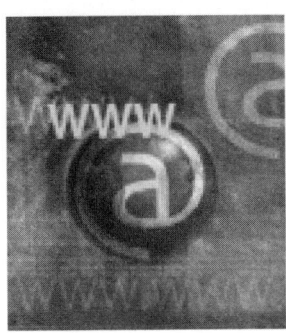

CHAPTER 1
THE NATURE OF ECONOMICS

LEARNING OBJECTIVES

After you have read this chapter, you should be able to

1. define economics;

2. distinguish between microeconomics and macroeconomics;

3. recognize the rationality assumption;

4. recognize elements of an economic model, or theory;

5. recognize that economics is ultimately concerned with human behavior;

6. define resource;

7. distinguish between positive economics and normative economics, and be able to classify specific statements under each category.

CHAPTER OUTLINE

1. Economics, a social science, is defined as the study of how people make choices to satisfy their wants.
   a. Wants are all the things that people would consume if they had unlimited income.
   b. Because wants are unlimited and people cannot satisfy all their wants, individuals are forced to make choices about how to spend their income and how to allocate their time.

2. Economics is broadly divided into microeconomics and macroeconomics.
   a. Microeconomics studies decision making undertaken by individuals (or households) and by firms.
   b. Macroeconomics studies the behavior of the economy taken as a whole; it deals with such economy-wide phenomena as unemployment, the price level, and national income.

3. Economists assume that individuals are motivated by self-interest and respond predictably to opportunities for gain.

a. The rationality assumption is that individuals act *as if* they were rational.
b. Self-interest often means a desire for material well-being, but it can also be defined broadly enough to incorporate goals relating to love, friendship, prestige, power, and other human characteristics.
c. By assuming that people act in a rational, self-interested way, economists can generate testable theories concerning human behavior.

4. Economics is a social science.
   a. Economists develop models, or theories, which are simplified representations of the real world.
   b. Such models help economists to understand, explain, and predict economic phenomena in the real world.
   c. Like other social scientists, economists usually do not perform laboratory experiments; they typically examine what already has occurred in order to test their theories.
   d. Economic theories, like all scientific theories, are simplifications—and in that sense they are "unrealistic."
   e. Economists, as do all scientists, employ assumptions; one important economic assumption is "all other things being equal."
   f. Models or theories are evaluated on their ability to predict, and not on the realism of the assumptions employed.
   g. Economic models relate to behavior, not thought processes.

5. Economists maintain that the unit of analysis is the individual; members of a group are assumed to pursue their own goals rather than the group's objectives.

6. Positive economics is objective and scientific in nature, and deals with testable *if this, then that* hypotheses.

7. Normative economics is subjective and deals with value judgments, or with what *ought* to be.

KEY TERMS
Aggregates
Ceteris paribus
 (other things being equal)
Economics

Empirical
Incentives
Macroeconomics
Microeconomics

Models (or theories)
Resource
Wants

KEY CONCEPTS
Normative economics
Positive economics

Rationality assumption

COMPLETION QUESTIONS
Fill in the blank, or circle the correct term.

1. Because it is impossible to have all that we want, people are forced to make _____.

2. Economics is a (natural, social) science.

3. Economics is the study of how people make _____ to satisfy their _____.

4. Microeconomics deals with (individual units, the whole economy).

5. A nation's unemployment level is analyzed in (microeconomics, macroeconomics).

6. (Macroeconomics, Microeconomics) studies the causes and effects of inflation.

7. Economists maintain that a member of a group usually attempts to make decisions that are in (her own, the group's) interest.

8. The rationality assumption is that individuals (believe, act as if) they are rational.

9. Economic models are (simplified, realistic) representations of the real world.

10. The *ceteris paribus* assumption enables economists to consider (one thing at a time, everything at once).

11. Married men are more likely to earn higher incomes if their wives work (inside, outside) the home.

12. Economists maintain that incentives (are, are not) important to decision making.

13. Economists define self-interest (narrowly, broadly).

14. Economists take the (individual, group) as the unit of analysis.

15. Economic statements that are testable and are of an "if/then" nature are (positive, normative).

TRUE-FALSE QUESTIONS
Circle the **T** if the statement is true, the **F** if it is false. Explain to yourself why a statement is false.

T F 1. Economics is the study of how people think about economic phenomena.

T F 2. The economists' definition of self-interest includes only the pursuit of material goods.

T F 3. Macroeconomics deals with aggregates, or totals, of economic variables.

T F 4. When economists attempt to predict the number of workers a firm will employ, they are studying macroeconomics.

T F 5. Economists maintain that people respond in a predictable way to economic incentives.

T F 6. The rationality assumption is that individuals attempt, quite consciously, to make rational economic decisions, and will admit to it.

T F 7. It is justifiable to criticize theories on the realism of the assumptions employed.

T F 8. Households cannot be thought of as producers.

T F 9. Because economics is a science, economists do not make normative statements.

T F 10. Medical issues such as what form of treatment to use for kidney dialysis patients are outside the scope of economics.

MULTIPLE CHOICE QUESTIONS
Circle the letter that corresponds to the best answer.

1. Economics is
   a. a natural science.
   b. nonscientific.
   c. a social science.
   d. usually studied through lab experiments.

2. Wants include desires for
   a. material possessions.
   b. love.
   c. power.
   d. All of the above

3. Which of the following areas of study is concerned, primarily, with microeconomics?
   a. the steel industry
   b. inflation
   c. the national unemployment rate
   d. national income determination

4. Macroeconomic analysis deals with
   a. the steel industry.
   b. how individuals respond to an increase in the price of gasoline.
   c. inflation.
   d. how a change in the price of energy affects a family.

5. Economists maintain that Mr. Smith will usually make decisions that promote the interests of
   a. his colleagues at work.
   b. himself.
   c. his class.
   d. his race.

6. Economic models
   a. use unrealistic assumptions.
   b. are seldom tested in laboratories.
   c. are concerned with how people behave, not with how they think.
   d. All of the above

7. An economic model is justifiably criticized if
   a. its assumptions are not realistic.
   b. it cannot be tested in a controlled, laboratory experiment.
   c. it fails to predict.
   d. All of the above

8. Which of the following is true of the "marriage premium"?
   a. It amounts to about 20 percent of women's earnings and about 10 percent of men's earnings.
   b. It results from one spouse handling the bulk of household management tasks.
   c. Both married men and married women earn equal marriage premiums.
   d. It is most commonly observed among women instead of men.

9. Economics
   a. is a natural science.
   b. is concerned with how people respond to incentives.
   c. is unconcerned with value judgments.
   d. deals with assumptions and therefore is unrealistic.

10. As is true of a road map showing how a traveler can move about a geographic region, a model of economic behavior typically
    a. omits trivial details and emphasizes factors most relevant to the problem under consideration.
    b. makes no simplifying assumptions, so that every feature of a problem is taken into account.
    c. must be rejected if it leaves out some information, even if it makes correct predictions.
    d. includes each and every element of a problem confronting an individual or group.

11. Which of the following is a normative economics statement?
    a. If price rises, people will buy less.
    b. If price rises, people will buy more.
    c. If price rises the poor will be injured; therefore price should not be permitted to rise.
    d. If price rises people will buy less; therefore we ought to observe that quantity demanded falls.

12. Which of the following is a positive economics statement?
    a. Full employment policies should be pursued.
    b. If minimum wage rates rise, then unemployment will rise.
    c. We should take from the rich and give to the poor.
    d. The government should help the homeless.

13. Normative economics statements
    a. are testable hypotheses.
    b. are value-free.
    c. are subjective, value judgments.
    d. can be scientifically established.

MATCHING
Choose the item in Column (2) that best matches an item in Column (1).

(1)	(2)
a. normative economics	f. nonscientific value judgments
b. macroeconomics	g. objective, scientific hypotheses
c. self-interest	h. study of individual behavior
d. positive economics	i. study of economic aggregates
e. microeconomics	j. rational behavior

ANSWERS TO CHAPTER 1

COMPLETION QUESTIONS

1. choices
2. social
3. choices; wants
4. individual units
5. macroeconomics
6. macroeconomics
7. her own
8. act as if
9. simplified
10. one thing at a time
11. inside
12. are
13. broadly
14. individual
15. positive

TRUE-FALSE QUESTIONS

1. F Economics is the study of how people make choices to satisfy their wants.
2. F Economists have a broader definition of self-interest; wants include power, friendship, love, and so on.
3. T
4. F The example is about microeconomics.
5. T
6. F That assumption is merely that people act *as if* they are rational.
7. F All theories employ unrealistic assumptions; what matters is how well they predict.
8. F Households can be thought of as combining goods and time to produce outputs such as meals.
9. F Economists, like other scientists, can and do make normative statements.
10. F Choices of all types, including among alternative medical treatments, can be subjected to economic analysis.

MULTIPLE CHOICE QUESTIONS

1.c;   2.d;   3.a;   4.c;   5.b;   6.d;   7.c;   8.b;   9.b;   10.a;
11.c;   12.b;   13.c.

MATCHING

a and f; b and i; c and j; d and g; e and h

GLOSSARY TO CHAPTER 1

Aggregates   Total amounts or quantities; aggregate demand, for example, is total planned expenditures throughout a nation.

***Ceteris paribus* assumption**   The assumption that nothing changes except the factor or factors being studied.

Economics   The study of how people allocate their limited resources to satisfy their unlimited wants.

Empirical   Relying on real-world data in evaluating the usefulness of a model.

Incentives   Rewards for engaging in a particular activity.

Macroeconomics The study of the behavior of the economy as a whole, including such economywide phenomena as unemployment, the price level, and national income.

Microeconomics The study of decision making undertaken by individuals (or households) and by firms.

Models, or theories Simplified representations of the real world used as the basis for predictions or explanations.

Normative economics Analysis involving value judgments about economic policies; relates to whether things are good or bad. A statement of what *ought to* be.

Positive economics Analysis that is *strictly* limited to making either purely descriptive statements or scientific predictions; for example, "If A, then B." A statement of *what is*.

Rationality assumption The assumption that people do not intentionally make decisions that would leave them worse off.

Resources Things used to produce other things to satisfy people's wants.

Wants What people would buy if their incomes were unlimited.

CHAPTER 2
SCARCITY AND THE WORLD OF TRADE-OFFS

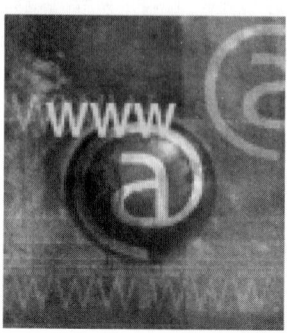

LEARNING OBJECTIVES

After you have studied this chapter, you should be able to

1. define production, scarcity, resources, land, labor, human and physical capital, entrepreneurship, goods, services, opportunity costs, production possibilities curve, technology, efficiency, inefficient point, law of increasing relative costs, specialization, absolute advantage, comparative advantage, and division of labor;

2. distinguish between a free good and an economic good;

3. determine the opportunity cost of an activity, when given sufficient information;

4. draw production possibilities curves under varying assumptions, and recognize efficient and inefficient points relating to such curves;

5. understand the difference between a person's or nation's absolute advantage and comparative advantage.

CHAPTER OUTLINE

1. Because individuals or communities do not have the resources to satisfy all their wants, scarcity exists.
   a. If society can get all that it wants of good A when the price of good A is zero, good A is not scarce.
   b. If the price of good B is zero, and society cannot get all that it wants of good B, then B is scarce.
   c. Because resources, or factors of production, are scarce, the outputs they produce are scarce.
      i. Land, the natural resource, includes all the gifts of nature.
      ii. Labor, the human resource, includes all productive contributions made by individuals who work.
      iii. Physical capital, the man-made resource, includes the machines, buildings, and tools used to produce other goods and services.
      iv. Human capital includes the education and training of workers.

v. Entrepreneurship includes the functions of organizing, managing, assembling, and risk-taking necessary for business ventures.
d. Goods include anything from which people derive satisfaction, or happiness.
   i. Economic goods are scarce.
   ii. Noneconomic goods are not scarce.
   iii. Services are intangible goods.
e. Economists distinguish between wants and needs; the latter are objectively undefinable.

2. Because of scarcity, choice and opportunity costs arise.
   a. Due to scarcity, people trade off options.
   b. The production possibilities curve (PPC) is a graph of the trade-offs inherent in a decision.
      i. When the amount of one resource or good that must be given up to produce an additional units of another resource or good remains constant, the PPC is a straight line.
      ii. When the amount of one resource or good that must be given up to produce an additional unit of another resource or good rises, the PPC is bowed outward.
      iii. A point on a PPC is an efficient point; points inside a PPC are inefficient; points outside the PPC are unattainable (impossible), by definition.
   c. People have an economic incentive to specialize in that endeavor for which they have a comparative advantage.
   d. The process of division of labor increases output and permits specialization.

3. Economic growth can be depicted through PPCs.
   a. There is a trade-off between present consumption and future consumption.
   b. If a nation produces fewer consumer goods and more capital goods now, then it can consume more goods in the future than would otherwise be the case.

4. Opportunity costs are important determinants of voter turnouts for elections.
   a. Instead of spending time going to and from a polling site and waiting in line to cast a ballot, an individual can be earning income or engaging in leisure activities, so every individual who casts a ballot incurs an opportunity cost of voting.
   b. Voter turnout at elections would be higher if society found ways to reduce the time commitment required to cast ballots.

KEY TERMS
Land
Labor
Physical capital
Human capital
Consumption
Economic goods
Goods
Services
Entrepreneurship
Technology
Production possibilities curve
Division of labor
Inefficient point
Scarcity
Production
Specialization

KEY CONCEPTS
Efficiency
Absolute advantage
Opportunity cost
Comparative advantage
Law of increasing relative cost

COMPLETION QUESTIONS
Fill in the blank, or circle the correct term.

1. The factors of production include _____, _____, _____, _____, and _____.

2. People tend to specialize in those activities for which they have (a comparative, an absolute) advantage.

3. When people choose jobs that maximize their income, they are specializing according to their _____ advantage.

4. If at a zero price quantity demanded exceeds quantity supplied for a good, that good is a(n) _____; if at a zero price quantity supplied exceeds quantity demanded for a good, that good is a(n) _____.

5. The _____ of good A is the highest-valued alternative that must be sacrificed to attain it.

6. If the opportunity cost of additional units of a good remains constant, the production possibilities curve will be (linear, bowed outward); if the opportunity cost of additional units of a good rises, the production possibilities curve will be (linear, bowed outward).

7. Because specialized resources are more suited to specific tasks, the opportunity cost of producing additional units of a specific good will (rise, fall).

8. If an economy is inefficient, its actual output combination will lie (inside, outside) the production possibilities curve.

TRUE-FALSE QUESTIONS
Circle the **T** if the statement is true, the **F** if it is false. Explain to yourself why a statement is false.

T F 1. Most individuals' needs exceed their wants.

T F 2. Because resources are scarce, the goods that they produce are also scarce.

T F 3. For most activities no opportunity cost exists.

T F 4. If a production possibilities curve is linear, the opportunity cost of producing additional units of a good rises.

T F 5. At any given moment in time, it is impossible for an economy to be inside its production possibilities curve.

T F 6. The cost to society of lowering the speed limit is zero.

T F 7. People have little incentive to specialize in jobs for which they have a comparative advantage.

T F 8. Economic growth shifts the production possibilities curve outward.

T F 9. If the price to a specific user is zero, the good must be a noneconomic good.

T F 10. The opportunity cost of casting ballots in elections is likely to be almost the same for all voters.

MULTIPLE CHOICE QUESTIONS
Circle the letter that corresponds to the best answer.

1. Because of scarcity
    a. people are forced to make choices.
    b. opportunity costs exist.
    c. people face trade-offs.
    d. All of the above

2. Which of the following is **NOT** considered to be "land"?
    a. bodies of water
    b. fertility of soil
    c. capital
    d. climate

3. Which of the following words does **NOT** belong with the others?
    a. opportunity cost
    b. economic bad
    c. scarcity
    d. economic good

4. Which statement concerning a production possibilities curve is **NOT** true?
    a. A trade-off exists along such a curve.
    b. It is usually linear.
    c. Points inside it indicate inefficiency.
    d. A point outside it is currently impossible to attain.

5. The production possibilities curve is bowed outward because
    a. the relative cost of producing a good rises.
    b. of the law of decreasing relative cost.
    c. all resources are equally suited to the production of any good.
    d. All of the above

6. When nations and individuals specialize,
    a. overall living standards rise.
    b. trade and exchange increase.
    c. people become more vulnerable to changes in tastes and technology.
    d. All of the above

7. When a nation expands its capital stock, it is usually true that
    a. it must forego output of some consumer goods in the present.
    b. the human capital stock must decline.
    c. fewer consumer goods will be available in the future.
    d. no opportunity cost exists for doing so.

8. Ms. Boulware is the best lawyer and the best secretary in town.
    a. She has a comparative advantage in both jobs.
    b. She has an absolute advantage in both jobs.
    c. She has a comparative advantage in being a secretary.
    d. All of the above

9. From 2:00 to 4:00 on a Thursday afternoon, Mr. Stapleton, a fast-food worker who earns the minimum wage, waits while his daughter is examined at a pediatrician's office. Ms. Black, a successful marketing consultant who normally charges a fee of $100 per hour and who recently has turned down several potential clients, spends exactly the same amount of time waiting for her own daughter to be examined. The pediatrician charges Mr. Stapleton $100 for the office visit. Ms. Black also pays the pediatrician $100. We may conclude that during this two-hour period,
   a. both parents incurred identical child-raising costs.
   b. from an economic standpoint, it was irrational for Ms. Black to wait while the pediatrician examined her child.
   c. Ms. Black incurred a higher child-raising cost, because she otherwise could have been earning consulting fees during this time.
   d. Mr. Stapleton incurred a higher child-raising cost, because he otherwise could have been looking for a higher-paying job during this time.

MATCHING
Choose the item in Column (2) that best matches an item in Column (1).

(1)

a. absolute advantage
b. efficiency
c. trade-offs
d. comparative advantage
e. resource
f. economic good
g. inefficiency
h. opportunity cost

(2)

i. production possibilities curve
j. specialization
k. capital
l. ability to produce at a lower unit cost
m. specializing in one's comparative advantage
n. society cannot get all it wants at at a zero price
o. highest-valued foregone alternative
p. inside PPC

WORKING WITH GRAPHS

1. Given the following information, graph the production possibilities curve in the space provided and then use the graph to answer the questions that follow.

Combination (points)	Autos (100,000 per year)	Wheat (100,000 tons per year)
A	16	0
B	14	4
C	12	7
D	9	10
E	5	12
F	0	13

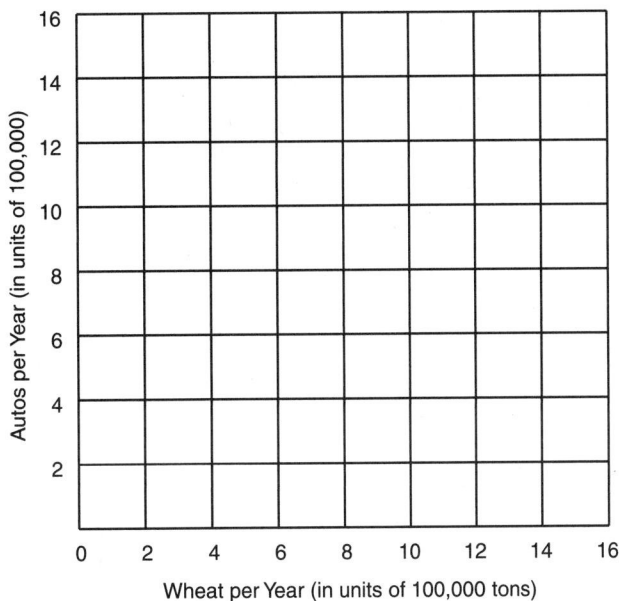

a. If the economy is currently operating at point C, what is the opportunity cost of moving to point D? to point B?
b. Suppose that the economy is currently producing 1,200,000 autos and 200,000 tons of wheat per year. Label this point in your graph with the letter G. At point G the economy would be suffering from what? At point G we can see that it is possible to produce more wheat without giving up any auto production, or produce more autos without giving up any wheat production, or produce more of both. Label this region in your graph. This region appears to contradict the definition of a production possibilities curve. What is the explanation for this result?
c. Suppose a new fertilizer compound is developed that will allow the economy to produce an additional 150,000 tons of wheat per year if no autos are produced. Sketch in a likely representation of the effect of this discovery, assuming all else remains constant.
d. What sort of impact (overall) will this discovery have on the opportunity cost of more wheat production at an arbitrary point on the new production possibilities curve, as compared to a point representing the same level of output of wheat on the original curve?

2. Consider the graphs below, then answer the questions that follow.

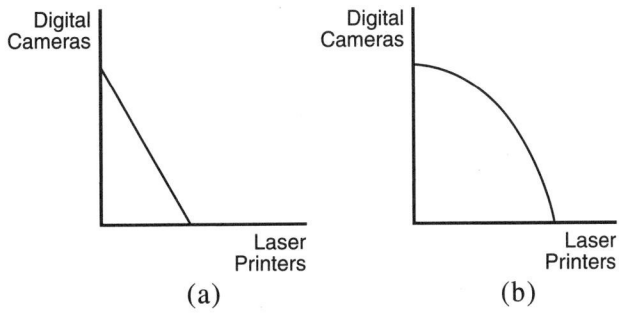

a. Which graph, a or b, shows constant relative costs of producing additional laser printers? Why?
b. Which graph, a or b, shows increasing relative costs of producing additional laser printers? Why?
c. Which graph seems more realistic, a or b? Why?

3. Graph the probable relationship between
   a. income and the amount of money spent on housing;
   b. annual rainfall in New York City and the annual value of ice cream sales in New Orleans;
   c. number of vegetarians per 10,000 people and meat sales per 10,000 people.

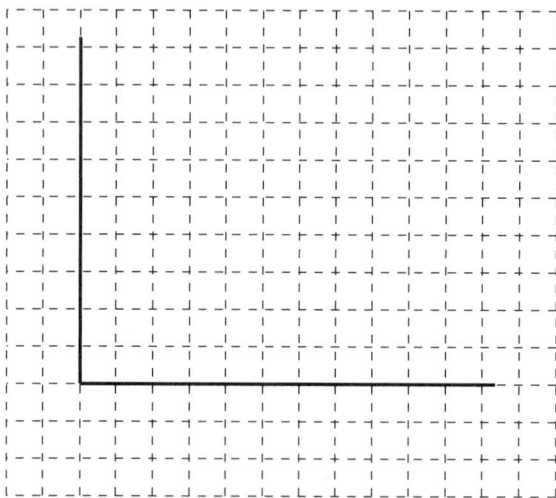

PROBLEMS

1. If a *nation* wants to increase its future consumption, it must forego some present consumption because it must allocate some resources to the production of capital goods. Suppose *you* want to increase your future consumption. Given a fixed lifetime income, what can you do?

2. Assume that Ms. Gentile values her time at $150 per hour because she has the opportunity to do consulting, and that Joe College values his time at $6 per hour. Assume that it costs $400 to fly from their hometown to San Francisco, and that the flight takes 6 hours. Assume that it costs $200 to take a bus, and that the bus trip takes 24 hours.
   a. What is the cheaper way to get to San Francisco for Ms. Gentile? Why?
   b. Which transportation is cheaper for Joe College? Why?

3. Suppose you have a friend currently working as a salesperson in a local computer store that sells personal computers. This friend is thinking about going back to school full-time to finish up work on her computer science degree. She explains to you that she earns $45,000 (after taxes) per year in her current job and that she estimates tuition will cost $4,800 per year. In addition she estimates fees, supplies, books, and miscellaneous expenses associated with attending school will run $2,400 per year. She wants to attend a university that is located directly across the street from the store where she currently works. She claims that she pays $1,050 per month for rent and utilities and that she spends about $600 per month on food, clothing, and related expenses.

   Using what you have learned, calculate and explain to your friend the opportunity cost to her of another year back at school.

4. The Hughes family consists of Mr. Hughes, Mrs. Hughes, and their son, Scotty. Assume that Mr. Hughes can earn $30 per hour (after taxes) any time he chooses, Mrs. Hughes can earn $5 per hour, and the family values homemaker activities at $6 per hour.
   a. Because the family requires income to purchase goods and services, who will probably work in the marketplace?
   b. Who will probably do the housework?
   c. If the family must pay $3 per hour to have its lawn mowed, who will be assigned that work?
   d. If Scotty can now earn $4 per hour on a job, who now might mow the grass?
   e. If wage rates in the marketplace for Mrs. Hughes rise to $7 per hour, what is the family likely to do?

ANSWERS TO CHAPTER 2

COMPLETION QUESTIONS

1. land, labor, physical capital, human capital, entrepreneurship
2. a comparative
3. comparative
4. economic good; economic "bad"
5. opportunity cost
6. linear; bowed outward
7. rise
8. inside

TRUE-FALSE QUESTIONS

1. F Wants vastly exceed needs (which can't be defined anyway) for everyone.
2. T
3. F An opportunity cost exists for all activities.
4. F A linear PPC implies a constant cost of production.
5. F All an economy need be is inefficient to be inside the PPC.
6. F If people drive more slowly, they suffer opportunity costs for their time.
7. F People can earn more income in jobs for which they have a comparative advantage.
8. T
9. F An individual user *may* pay a zero price for a good, but that doesn't necessarily mean it's a noneconomic good.
10. F A fundamental component of the opportunity cost of casting ballots is wages that a voter foregoes by devoting time to this activity, and wages vary considerably across individuals.

MULTIPLE CHOICE QUESTIONS

1.d;   2.c;   3.b;   4.b;   5.a;   6.d;   7.a;   8.b;   9.c.

MATCHING

a and l; b and m; c and i; d and j; e and k; f and n; g and p; h and o

16 CHAPTER 2: SCARCITY AND THE WORLD OF TRADE-OFFS

WORKING WITH GRAPHS

1. See the following graph.

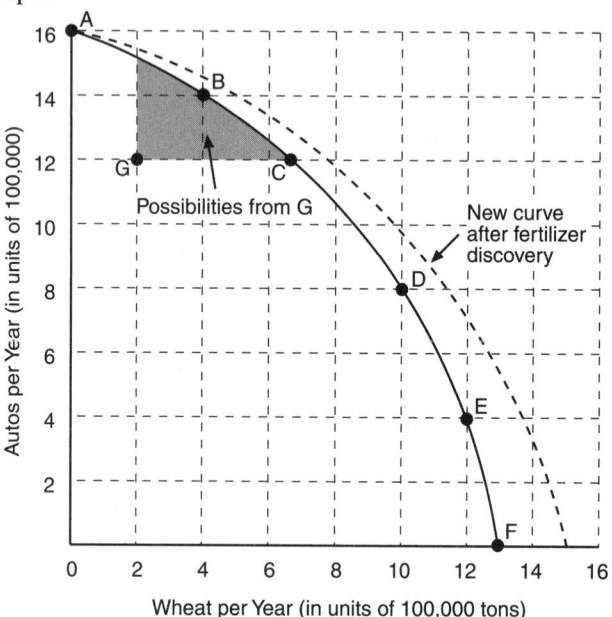

   a. The move from C to D "costs" 300,000 autos—that is, the economy must give up 300,000 autos (1,200,000 - 900,000) to make such a move. The move from C to B "costs" 300,000 tons of wheat. Notice that in both cases there are gains (C to D involves 300,000 more tons of wheat, and C to B means 200,000 more autos are produced), but we measure opportunity costs in terms of movements along a production possibilities curve and what has to be given up to make the choice reflected in the move.
   b. See the preceding graph. Remember, the production possibilities curve shows all possible combinations of two goods that an economy can produce by the efficient use of all available resources in a specified period of time. Since point G is not on the production possibilities curve, the statement contained in this portion of the question does not contradict the definition of the curve. Point G is inside the curve, which implies available resources are not being used efficiently.
   c. See the preceding graph.
   d. It will lower the opportunity cost of additional wheat production.

2. a. Graph a shows constant relative costs because the PPC is linear.
   b. Graph b shows increasing costs because the PPC is bowed out.
   c. Graph b is more realistic because it is likely that the production of laser printers and digital cameras requires specialized resources.

3. a. The graph should be upward sloping from left to right.
   b. There should be no systematic relationship between these two variables.
   c. The graph should be downward sloping from left to right.

PROBLEMS

1. If you want to increase your future consumption—for retirement, say—then you will have to save more out of your current income. The principal and interest that accrue will permit you to purchase more goods in the future than you otherwise would have been able to. Note that by doing so you—as an individual—must forego some present consumption in order to increase your future consumption. In that sense, what is true for the nation is also true for an individual.

2. a. plane; flying costs her $400 plus 6 hours times $150 per hour, or $1,300, while taking a bus would cost her $200 plus 24 hours times $150 per hour, or $3,800.
   b. bus; taking the bus costs him a total of $344, while his total cost of flying is $436.

3. The opportunity cost of another year back at school for your friend is as follows:

Foregone after-tax salary	$45,000
Tuition costs	4,800
Expenses associated with school	2,400
Total opportunity costs	$52,200

4. a. Mr. Hughes
   b. Mrs. Hughes or Scotty
   c. Scotty
   d. The family (or perhaps Scotty) will hire someone to mow the lawn.
   e. Mrs. Hughes may enter the labor force and the family may hire someone to do housework.

GLOSSARY TO CHAPTER 2

Absolute advantage The ability to produce more units of a good or service using a given quantity of labor or resource inputs. Equivalently, the ability to produce the same quantity of a good or service using fewer units of labor or resource inputs.

Comparative advantage The ability to produce a good or service at a lower opportunity cost compared to other producers.

Consumption The use of goods or services for personal satisfaction.

Division of labor The segregation of a resource into different specific tasks; for example, one automobile worker puts on bumpers, another doors, and so on.

Economic goods Goods that are scarce, for which the quantity demanded exceeds the quantity supplied at a zero price.

Efficiency The case in which a given level of inputs is used to produce the maximum output possible. Alternatively, the situation in which a given output is produced at minimum cost.

Entrepreneurship The factor of production involving human resources that perform the functions of raising capital, organizing, managing, assembling other factors of production, and making basic business policy decisions. The entrepreneur is a risk taker.

Goods All things from which individuals derive satisfaction or happiness.

Human capital The accumulated training and education of workers.

Inefficient point Any point below the production possibilities frontier, at which resources are being used inefficiently.

Labor Productive contributions of humans who work, involving both mental and physical activities.

Land The natural resources that are available from nature. Land as a resource includes location, original fertility and mineral deposits, topography, climate, water, and vegetation.

Law of increasing relative cost The observation that the opportunity cost of additional units of a good generally increases as society attempts to produce more of that good. This accounts for the bowed-out shape of the production possibilities curve.

Opportunity cost The highest-valued, next-best alternative that must be sacrificed to attain something or to satisfy a want.

Physical capital All manufactured resources, including buildings, equipment, machines, and improvements to land that is used for production.

Production Any activity that results in the conversion of resources into products that can be used in consumption.

Production possibilities curve (PPC) A curve representing all possible combinations of total output that could be produced assuming (1) a fixed amount of productive resources of a given quality and (b) the efficient use of those resources.

Scarcity A situation in which the ingredients for producing the things that people desire are insufficient to satisfy all wants.

Services Mental or physical labor or help purchased by consumers. Examples are the assistance of physicians, lawyers, dentists, repair personnel, housecleaners, educators, retailers, and wholesalers; things purchased or used by consumers that do not have physical characteristics.

Specialization The division of productive activities among persons and regions so that no one individual or one area is totally self-sufficient. An individual may specialize, for example, in law or medicine. A nation may specialize in the production of coffee, computers, or cameras.

Technology Society's pool of applied knowledge concerning how goods and services can be produced.

CHAPTER 3: DEMAND AND SUPPLY

KEY TERMS
Demand schedule
Demand curve
Supply schedule
Supply curve
Market price
Market demand
Market
Money price

KEY CONCEPTS
Relative price
Law of demand
Normal goods
Inferior goods
Substitutes
Subsidy
Complements
Law of supply
Equilibrium
Shortage, or excess quantity demanded
Surplus, or excess quantity supplied

COMPLETION QUESTIONS
Fill in the blank, or circle the correct term.

1. A(n) _____ relates various possible prices to the quantities demanded at each price, and a(n) _____ relates various prices to the quantities supplied at each price.

2. A change in quantity demanded is a (movement along, shift in) the demand curve; and a change in demand is a(n) _____ the demand curve.

3. At the intersection of the supply and demand curves, the quantity supplied equals the quantity demanded, and at that price a(n) _____ exists; at a price above that intersection, quantity supplied exceeds quantity demanded and a(n) _____ exists; at a price below that intersection, quantity demanded exceeds quantity supplied, and a(n) _____ exists.

4. The law of demand states that, other things being equal, more is bought at a (lower, higher) price and less is bought at a(n) _____ price.

5. There is (a direct, an inverse) relationship between price and quantity demanded, and demand curves will be (positively, negatively) sloped.

6. When the other determinants of demand change, the entire demand curve shifts; the five major *ceteris paribus* conditions affecting demand are _____, _____, _____, _____, and _____.

7. If the demand for pizza rises, given the supply, then the equilibrium price of pizza will (rise, fall) and the equilibrium quantity will _____.

8. The law of supply relates prices to quantities supplied; in general, as price rises, quantity supplied _____. Therefore (a direct, an inverse) relationship exists, and the supply curve is (positively, negatively) sloped.

9. The supply curve is positively sloped because as price rises, producers have an incentive to produce (less, more).

10. When the determinants of supply change, the entire supply curve will shift; five major determinants of supply are _____, _____, _____, _____, and _____.

11. Videocassettes and videocassette players are (substitutes, complements); if the price of videocassette players rises, then the demand for videocassettes will _____.

12. When the price of peaches rises, the demand for pears rises; peaches and pears are (substitutes, complements).

13. *Analogy*: An excess quantity supplied is to a surplus as a(n) _____ is to a shortage.

14. A rise in demand causes the demand curve to shift to the (left, right); an increase in quantity demanded occurs when there is a movement (up, down) the demand curve.

15. By convention, economists plot (price, quantity) on the vertical axis and (price, quantity) on the horizontal axis.

TRUE-FALSE QUESTIONS
Circle the **T** if the statement is true, the **F** if it is false. Explain to yourself why a statement is false.

T F 1. A demand schedule relates quantity demanded to quantity supplied, other things being constant.

T F 2. A change in the quantity demanded of cigarettes results from a change in the price of cigarettes.

T F 3. A graphical representation of a demand curve is called a demand schedule.

T F 4. An increase in price leads to a leftward shift in demand and a rightward shift in supply.

T F 5. An increase in the price of CDs causes a rise in the supply of CDs.

T F 6. Buyers are concerned with absolute, not relative, prices.

T F 7. As producers increase output in the short run, the cost of additional units of output tends to rise.

T F 8. If the price of tennis racquets rises, the demand for tennis balls will tend to rise also.

T F 9. If the price of butter rises, the demand for margarine will rise.

T F 10. If price is below the equilibrium price, a shortage exists.

MULTIPLE CHOICE QUESTIONS

Circle the letter that corresponds to the best answer.

1. A demand schedule
   a. relates price to quantity supplied.
   b. when graphed, is a demand curve.
   c. cannot change.
   d. shows a direct relationship between price and quantity demanded.

2. If the price of milk rises, other things being constant,
   a. buyers will drink less milk.
   b. buyers will substitute milk for other beverages.
   c. the demand for milk will fall.
   d. the demand for cola drinks will fall.

3. Which of the following will **NOT** occur if the price of hamburger meat falls, other things being constant?
   a. The demand for hamburger buns will increase.
   b. People will substitute hamburgers for hot dogs.
   c. The demand for hot dogs will rise.
   d. The quantity of hamburgers demanded will increase.

4. If the price of good A rises and the demand for good B rises, then A and B are
   a. substitutes.
   b. complements.
   c. not related goods.
   d. not scarce goods.

5. Several years ago some cities in North Carolina passed a law that limited showers to 4 minutes, with a possible 30-day jail sentence for violators. Which of the following statements is probably true for those cities?
   a. A surplus of water existed.
   b. The price of water was too high.
   c. A shortage of water would exist regardless of how high its price got.
   d. The price of water was below the equilibrium price.

6. If the supply of gasoline rises, with a given demand, then
   a. the relative price of gasoline will rise.
   b. the equilibrium price of gasoline will rise.
   c. the equilibrium quantity of gasoline will increase.
   d. the equilibrium price and equilibrium quantity of gasoline will increase.

7. If income falls and the demand for steak falls, then steak is a(n)
   a. substitute good.
   b. complement good.
   c. normal good.
   d. inferior good.

Consider the graphs below when answering questions 8 and 9.

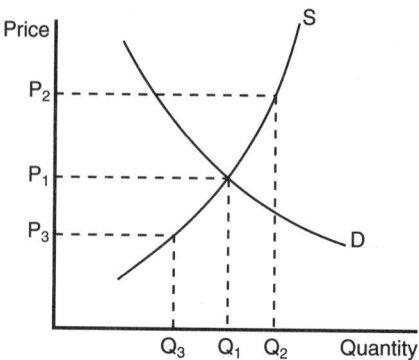

8. Given the figure above,
   a. the equilibrium price is P_1, and the equilibrium quantity is P_2.
   b. the equilibrium quantity is P_1.
   c. the equilibrium price is P_3, and the equilibrium quantity is Q_1.
   d. the equilibrium quantity is Q_1, and the equilibrium price is P_1.

9. Which of the following is **NOT** true?
   a. A shortage exists at P_2.
   b. The equilibrium price is P_1.
   c. An excess quantity demanded exists at P_3.
   d. The market-clearing price is P_1.

10. If the demand for hamburgers rises, with a given supply, then
    a. the supply of hamburgers will rise because price rises.
    b. the equilibrium price of hamburgers will fall and the equilibrium quantity will rise.
    c. the equilibrium quantity and the equilibrium price of hamburgers will rise.
    d. the quantity supplied of hamburgers will decrease.

11. If a shortage exists at some price, then
    a. sellers can sell all they desire to sell at that price.
    b. sellers have an incentive to raise the price.
    c. buyers cannot get all they want at that price.
    d. All of the above

12. Which of the following will lead to a rise in supply?
    a. an increase in the price of the good in question.
    b. a technological improvement in the production of the good in question.
    c. an increase in the price of labor used to produce the good in question.
    d. All of the above

13. Which of the following probably will **NOT** lead to a fall in the demand for hamburgers?
    a. a decrease in income.
    b. an expectation that the price of hamburgers will rise in the future.
    c. a decrease in the price of hot dogs.
    d. a change in tastes away from hamburgers.

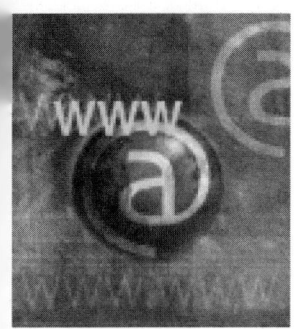

CHAPTER 3
DEMAND AND SUPPLY

LEARNING OBJECTIVES

After you have studied this chapter, you should be able to

1. define demand schedule, quantity demanded, supply schedule, quantity supplied, equilibrium, shortage, and surplus;

2. graph demand and supply curves from demand and supply schedules;

3. state the law of demand and state two reasons why we observe the law of demand;

4. enumerate five *ceteris paribus* conditions determining demand and five *ceteris paribus* conditions determining supply;

5. predict the effects of a change in the price of one good on the demand for (a) a substitute good, and (b) a complementary good;

6. recognize, from graphs, the difference between a change in demand and a change in quantity demanded, and the difference between a change in supply and a change in quantity supplied;

7. determine from a supply curve and a demand curve what the equilibrium price and the equilibrium quantity will be;

8. explain how markets eliminate surpluses and shortages.

CHAPTER OUTLINE

1. The law of demand states that at higher prices a lower quantity will be demanded than at lower prices, other things being equal.
   a. For simplicity, things other than the price of the good itself are held constant.
   b. Buyers respond to changes in relative, not absolute, prices.

2. The demand schedule for a good is a set of pairs of numbers showing various possible prices and the quantity demanded at each price, for some time period.
   a. Demand must be conceived of as being measured in constant-quality units.
   b. A demand curve is a graphic representation of the demand schedule and it is negatively sloped, reflecting the law of demand.
   c. A market demand curve for a particular good or service is derived by summing all the individual demand curves for that product.

3. The determinants of demand include all factors (other than the good's own price) that influence the quantity purchased.
   a. When deriving a demand curve, other determinants of demand are held constant. When such *ceteris paribus* conditions affecting demand do change, the original demand curve shifts to the left or to the right.
   b. The major determinants of demand are consumers' income, tastes and preferences; changes in their expectations about future relative prices; the price of substitutes and complements for the good in question; and number of buyers.
   c. A change in demand is a shift in the demand curve, whereas a change in quantity demanded is a movement along a given demand curve.

4. Supply is the relationship between price and the quantity supplied, other things being equal.
   a. The law of supply posits generally a direct, or positive, relationship between price and quantity supplied.
      i. As the relative price of a good rises, producers have an incentive to produce more of it.
      ii. As a firm produces greater quantities in the short run, a firm often requires a higher relative price before it will increase output.
   b. A supply schedule is a set of numbers showing prices and the quantity supplied at those various prices.
   c. A supply curve is the graphic representation of the supply schedule; it is positively sloped.
   d. By summing individual supply curves for a particular good or service we derive that good or service's market supply curve.
   e. The major determinants of supply are the prices of resources (inputs) used to produce the product; technology; taxes and subsidies; price expectations of producers; and the number of firms in an industry.
   f. Any change in the determinants of supply (listed in part e) causes a change in supply and therefore leads to a shift in the supply curve.
   g. A change in price, holding the determinants of supply constant, causes a movement along—but not a shift in—the supply curve.

5. By graphing demand and supply on the same coordinate system, we can find equilibrium at the intersection of the two curves.
   a. Equilibrium is a situation in which the plans of buyers and of sellers exactly coincide, so that there is neither excess quantity supplied nor excess quantity demanded; at the equilibrium price, quantity supplied equals quantity demanded.
   b. At a price below the equilibrium price, quantity demanded exceeds quantity supplied, and *excess quantity demanded*, or a shortage, exists.
   c. At a price above the equilibrium price, quantity supplied exceeds quantity demanded, and an *excess quantity supplied*, or a surplus, exists.
   d. Seller competition forces price down and eliminates a surplus.
   e. Buyer competition forces price up and eliminates a shortage.

14. When a demand curve is derived,
    a. quantity is in constant-quality units.
    b. the price of the good is held constant.
    c. money income changes.
    d. consumer tastes change.

15. If a surplus exists at some price, then
    a. sellers have an incentive to raise the price.
    b. buyers have an incentive to offer a higher price.
    c. sellers cannot sell all they wish to at that price.
    d. seller inventories are falling.

MATCHING
Choose the item in Column (2) that best matches an item in Column (1).

(1)	(2)
a. excess quantity demanded	k. relation between price and quantity demanded
b. supply curve	l. law of supply
c. demand curve	m. population increases
d. bread and butter	n. raw material prices rise
e. eyeglasses and contact lenses	o. community money income falls
f. demand shifts to the left	p. market clearing price
g. supply shifts to the left	q. surplus
h. equilibrium price	r. complements
i. equilibrium quantity rises	s. substitutes
j. excess quantity supplied	t. shortage

WORKING WITH GRAPHS

1. Use the demand schedule below to plot the demand curve on the following coordinate system. Be sure to label each axis correctly.

Price per bottle of shampoo	Quantity demanded of bottles of shampoo per week (in thousands)
$6	8
$5	10
$4	12
$3	14
$2	16
$1	18

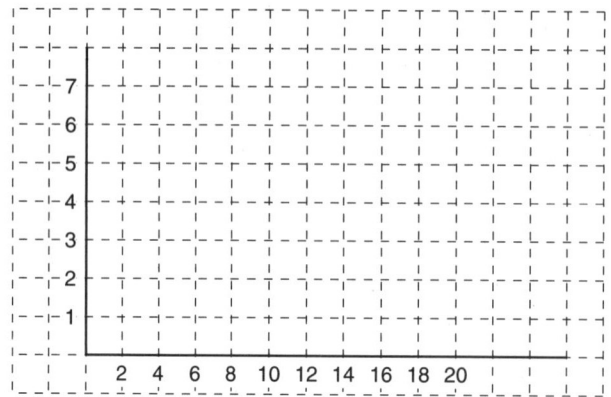

2. Use the supply schedule below to plot the supply curve on the coordinate system in problem 1.

Price per bottle of shampoo	Quantity supplied of bottles of shampoo per week (in thousands)
$6	18
$5	15
$4	12
$3	9
$2	6
$1	3

3. Using the graphs from problems 1 and 2, indicate on the graph the equilibrium price and the equilibrium quantity for bottles of shampoo. What is the equilibrium price? The equilibrium quantity?

4. Continuing with the same example, assume that the government mandates that shampoo cannot be sold for more that $3 per bottle. What is the quantity demanded at that price? The quantity supplied? Does a surplus or a shortage exist at that price?

5. Consider the two graphs below, in panels a and b. Which panel shows an increase in quantity demanded? Which shows a rise in demand?

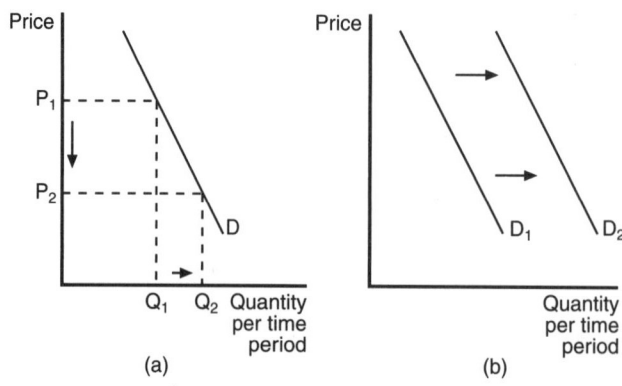

6. Distinguish between a fall in supply and a decrease in quantity supplied, graphically, using the space below. Use two panels.

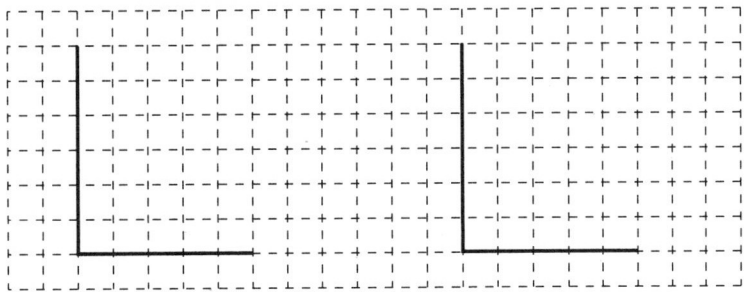

7. Consider the graphs below in panel (a). Then show, in panel (b), the new equilibrium price (label it P_1) and the new equilibrium quantity (label it Q_1) that result due to a change in tastes in favor of the good in question.

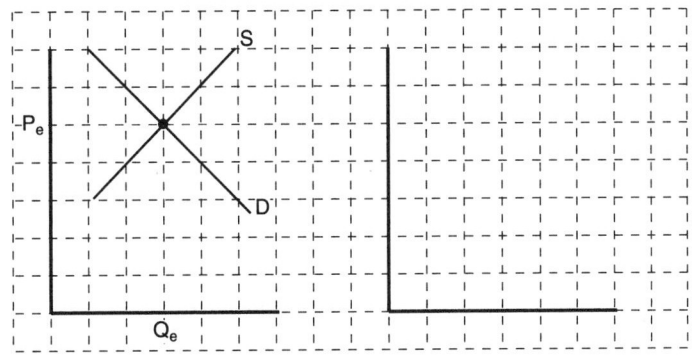

PROBLEMS

1. In the table below, monthly demand schedules for turkey are indicated. Assume that column 2 represents quantities demanded in October, column 3 represents November, and column 4 represents December.

(1) Price per pound	(2) Q_d	(3) Q_d	(4) Q_d
5 cents	10,000		16,000
10 cents	8,000		14,000
15 cents	6,000		12,000
20 cents	4,000		10,000
25 cents	2,000		8,000

   Fill in column 3 yourself. What happens to the demand for turkey in November relative to October and December? Why?

2. List the kinds of changes in the *ceteris paribus* conditions of demand that will lead to a decrease in demand. Be specific.

28 CHAPTER 3: DEMAND AND SUPPLY

3. List the kinds of changes in the *ceteris paribus* conditions of supply that will lead to a rise in supply. Be specific.

4. Concerning oranges, indicate whether each event leads to (i) a rightward shift in demand, (ii) a leftward shift in demand, (iii) an increase in quantity demanded, (iv) a decrease in quantity demanded, (v) a rightward shift in supply, (vi) a leftward shift in supply, (vii) an increase in quantity supplied, (viii) a decrease in quantity supplied.

   (Note: some events may lead to more than one of the above.)

   _____ a) An early frost in Florida destroys some orange groves.
   _____ b) Migrant workers organize a union and raise wage rates.
   _____ c) The price of oranges rises.
   _____ d) The price of oranges falls.
   _____ e) The Federal government lowers the price of oranges below equilibrium and freezes the price at the lower level.
   _____ f) Orange growers leave the industry.
   _____ g) Lemons (but not oranges) are demonstrated to cause cancer in lab rats.
   _____ h) The government subsidizes orange growers at 3 cents per orange.
   _____ i) News is released that the government forecast is for a poor orange crop this year.
   _____ j) The price of lemons rises (assume now that orange growers can also grow lemons).

5. For each of the statements (a) through (j) in the previous question, decide whether the market clearing *(equilibrium)* price will rise, fall, or be unaffected.

   a) _____
   b) _____
   c) _____
   d) _____
   e) _____
   f) _____
   g) _____
   h) _____
   i) _____
   j) _____

ANSWERS TO CHAPTER 3

COMPLETION QUESTIONS

1. demand curve or schedule; supply curve or schedule
2. movement along; shift in
3. equilibrium; surplus; shortage
4. lower; higher
5. inverse; negatively
6. income, tastes and preferences, prices of related goods, expectations about future relative prices, number of buyers
7. rise; rise
8. increases; direct; positively
9. more
10. prices of inputs, technology, taxes and subsidies, price expectations, number of firms in industry
11. complements; fall
12. substitutes
13. excess quantity demanded
14. right; down
15. price; quantity

TRUE-FALSE QUESTIONS

1. F   A demand schedule relates quantity demanded to price.
2. T
3. F   A graphical representation of a demand schedule is a demand curve.
4. F   An increase in price leads to a decrease in quantity demanded and an increase in quantity supplied.
5. F   It increases the quantity of CDs supplied, which is a movement along the supply curve, not a shift in the curve.
6. F   Buyers respond to changes in relative prices.
7. T
8. F   The demand for tennis balls will tend to *fall*, because they are complements.
9. T
10. T

MULTIPLE CHOICE QUESTIONS

1.b;   2.a;   3.c;   4.a;   5.d;   6.c;   7.c;   8.d;   9.a;   10.c;
11.d;   12.b;   13.b;   14.a;   15.c.

MATCHING

a and t;   b and l;   c and k;   d and r;   e and s;   f and o;   g and n;   h and p;
i and m;   j and q

WORKING WITH GRAPHS

1. See graphs.
2. See graphs.

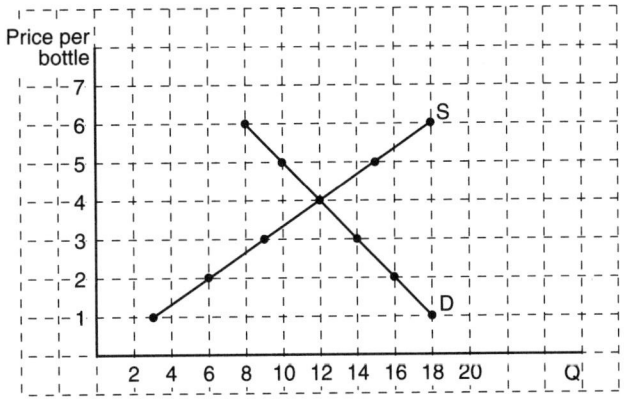

3. $4; 12,000 bottles
4. 14,000 bottles; 9,000 bottles; shortage
5. Panel (a); Panel (b)

6.

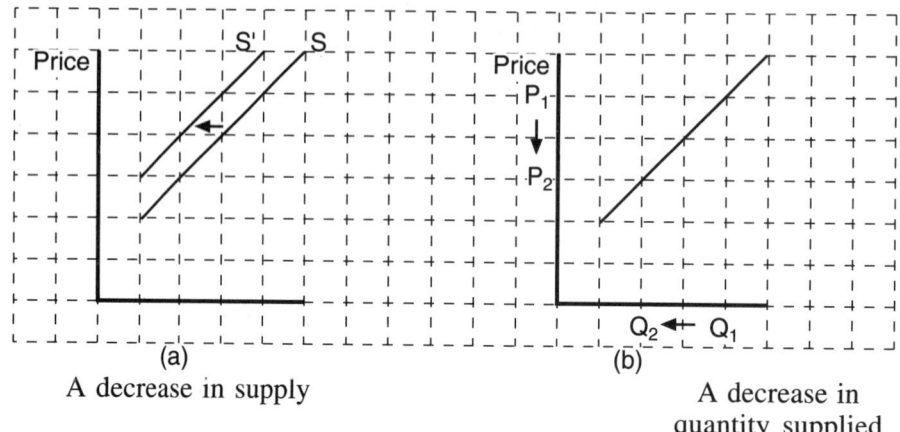

(a) A decrease in supply

(b) A decrease in quantity supplied

7.

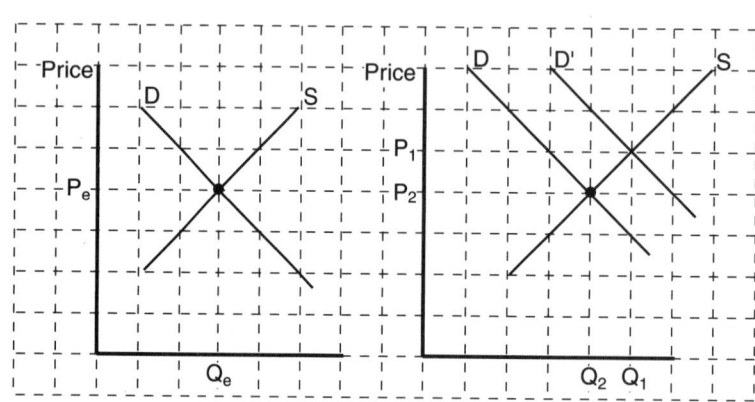

PROBLEMS

1. Due to Thanksgiving, rises significantly relative to October, and (perhaps) slightly relative to December.
2. Income falls for a normal good; change in tastes occurs away from the good; price of a substitute falls or price of a complement rises; expectations exist that the good's future relative price will fall; decrease in the number of buyers occurs.
3. Reduction in the price of inputs; technological advances; decrease in a sales tax on the good or increase in the (per unit) subsidy of the good; expectation that the future relative price will fall; increase in the number of firms in the industry.

4.
a. vi and iv
b. vi and iv
c. iv and vii
d. iii and viii
e. iii and viii
f. vi and iv
g. i and vii
h. v and iii
i. i
j. i and vi

5.
a. rise
b. rise
c. be unaffected
d. be unaffected
e. be unaffected
f. rise
g. rise
h. fall
i. rise
j. rise

GLOSSARY TO CHAPTER 3

***Ceteris paribus* conditions** Determinants of the relationship between price and quantity that are unchanged along a curve; changes in these factors cause the curve to shift.

Complements Two goods are complements if both are used together for consumption or enjoyment—for example, coffee and cream. The more you buy of one, the more you buy of the other. For complements, a change in the price of one causes an opposite shift in the demand for the other.

Demand A schedule of how much of a good or service people will purchase at any price during a specified time period, other things being constant.

Demand curve A graphical representation of the demand schedule; a negatively sloped line showing the inverse relationship between the price and the quantity demanded (other things being equal).

Equilibrium The situation when quantity supplied equals quantity demanded at a particular price.

Inferior goods Goods for which demand falls as income rises.

Law of demand The observation that there is a negative, or inverse, relationship between the price of any good or service and the quantity demanded, holding other factors constant.

Law of supply The observation that the higher the price of a good, the more of that good sellers will make available over a specified time period, other things being equal.

Market All of the arrangements that individuals have for exchanging with one another. Thus we can speak of the labor market, the automobile market, and the credit market.

Market clearing, or equilibrium, price The price that clears the market, at which quantity demanded equals quantity supplied; the price where the demand curve intersects the supply curve.

Market demand The demand of all consumers in the marketplace for a particular good or service. The summing at each price of the quantity demanded by each individual.

Money price That price that we observe today in terms of today's dollars; also called the *absolute*, *nominal*, or *current* price.

Normal goods Goods for which demand rises as income rises. Most goods are normal goods.

Relative price The price of one commodity divided by the price of another commodity; the number of units of one commodity that must be sacrificed to purchase one unit of another commodity.

Shortage A situation in which quantity demanded is greater than quantity supplied at a price below the market clearing price.

Subsidy A negative tax; a payment to a producer from the government, usually in the form of a cash grant.

Substitutes Two goods are substitutes when either one can be used for consumption to satisfy a similar want—for example, coffee and tea. The more you buy of one, the less you buy of the other. For substitutes, the change in the price of one causes a shift in demand for the other in the same direction as the price change.

Supply A schedule showing the relationship between price and quantity supplied for a specified period of time, other things being equal.

Supply curve The graphical representation of the supply schedule; a line (curve) showing the supply schedule, which generally slopes upward (has a positive slope), other things being equal.

Surplus A situation in which quantity supplied is greater than quantity demanded at a price above the market clearing price.

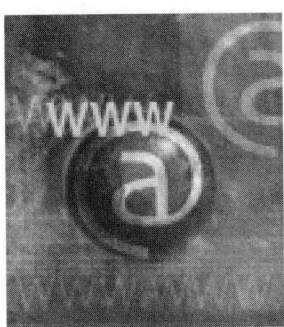

CHAPTER 4
EXTENSIONS OF DEMAND AND SUPPLY ANALYSIS

LEARNING OBJECTIVES

After you have studied this chapter, you should be able to

1. define price system, voluntary exchange, terms of exchange, transactions costs, price controls, price ceiling, price floor, nonprice rationing devices, black market, rent control, minimum wage, and import quota;

2. predict what happens to equilibrium price and equilibrium quantity when supply increases or decreases relative to demand, and when demand increases or decreases relative to supply;

3. predict what happens to the relative price of a good or resource if it becomes more or less scarce;

4. differentiate between the causes of short-run and long-run (prolonged) shortages;

5. recognize various methods of rationing goods and services;

6. recognize, from graphs, how a black market emerges;

7. enumerate several consequences of rent control;

8. recognize several consequences of government quantity restrictions;

9. recognize the consequences of price floors and the causes of prolonged surpluses.

CHAPTER OUTLINE

1. In a price system (free enterprise) voluntary exchange typically determines price; buyers and sellers transact with a minimum amount of governmental interference.
   a. Under a system of voluntary exchange, the terms of exchange (the terms, usually price, under which trade takes place) are set by the forces of supply and demand.
   b. Markets reduce transactions costs (all the costs associated with exchanging, including such costs associated with gathering information and enforcing contracts).

c. Under voluntary exchange *both* buyers and sellers are presumed to benefit—otherwise the transactions would not continue.

2. Changes in demand and/or supply lead to changes in the equilibrium price and the equilibrium quantity.
   a. If demand shifts to the right (left), given supply, then the equilibrium price rises (falls) and the equilibrium quantity rises (falls).
   b. If supply shifts to the right (left), given demand, then the equilibrium price falls (rises) and the equilibrium quantity rises (falls).
   c. When both supply and demand change, it is not always possible to predict the effects on the equilibrium price and the equilibrium quantity.

3. Prices are not always perfectly flexible.
   a. If prices are inflexible, published prices will not change very much, but such hidden price rises as a quality reduction might occur.
   b. Markets do not always move to equilibrium (given a change in demand or supply) immediately; hence shortages can emerge in the short run.

4. Price reflects relative scarcity and performs a rationing function.
   a. If an input or output becomes less scarce (more scarce), its relative price will fall (rise).
   b. If governments prevent prices from rising to their equilibrium level, via a price control or ceiling, then goods cannot (legally) be allocated to the highest bidders and prolonged shortages result; other forms of rationing emerge.
   c. During prolonged shortages, such nonprice rationing devices as cheating, long lines, first-come first-served, political power, physical force, and other nonmarket forces arise.
   d. Governments also interfere in markets by putting price floors on price; for example, governments impose minimum wage rates, and they have put price floors on agricultural goods, which have caused surpluses.

5. Rent controls are governmentally imposed price ceilings on rental apartments, which lead to predictable results; nonprice rationing for apartments results.

6. The government has put price floors in several markets.
   a. For many years, price supports created minimum prices for agricultural goods.
   b. When the government sets minimum wages above the equilibrium, some unemployment is created.
   c. Governments sometimes restrict quantity directly through import quotas, which prohibit the importation of more than a specified quantity of a particular good in a one-year period.

KEY TERMS
Price ceiling
Transactions costs

Price floor
Minimum wage

Black market
Import quota

KEY CONCEPTS
Price controls
Voluntary exchange

Rent control
Terms of exchange

Price system
Nonprice rationing devices

COMPLETION QUESTIONS
Fill in the blank, or circle the correct term.

1. Resources are scarce; therefore we cannot have all we want at a (zero, positive) price and there will be various ways in which people will _____ for resources.

2. If demand shifts to the left, given supply, then the equilibrium price will (rise, fall) and the equilibrium quantity will _____.

3. If supply shifts to the right, given demand, then the equilibrium price will _____ and the equilibrium quantity will _____.

4. If both demand and supply shift to the right, then the equilibrium price (will rise, will fall, is indeterminate) and the equilibrium quantity (will rise, will fall, is indeterminate).

5. If both demand and supply shift to the left, then the equilibrium price (will rise, will fall, is indeterminate), and the equilibrium quantity (will rise, will fall, is indeterminate).

6. If the demand for good A or resource A rises relative to its supply, A has become relatively (less scarce, more scarce) and its relative price will (rise, fall); if the demand for good B or resource B falls relative to its supply, then B has become relatively _____, and its relative price will _____.

7. If the published price of good A remains constant, but its quality falls, then its relative price has actually (risen, fallen). If the published price of good A remains constant, but people have to wait in line to get it, then the relative price of good A has actually _____, because people have an opportunity cost for their _____.

8. If the demand for a good rises relative to its supply, that good becomes (less scarce, more scarce) and its relative price will (rise, fall); this leads to (a decrease, an increase) in the quantity of the good supplied by producers of the item.

9. Price performs a(n) _____ function; inputs or outputs go to the _____ bidders, if people are free to exchange voluntarily in markets. If such economic freedoms do not exist, then other (price, nonprice) determinants will allocate goods and services.

10. Price controls that put a price ceiling on goods and services create (surpluses, shortages); and price floors create (surpluses, shortages).

11. If governments place price (floors, ceilings) on goods, then black markets might emerge.

12. Rent control is a form of price (floor, ceiling); rent control (increases, reduces) the future supply of apartment construction, (increases, reduces) tenant mobility, (improves, causes a deterioration in) the quality of the existing stock of apartments, and hurts _____.

13. By prohibiting the sale and use of tobacco products, the government would cause the supply of cigarettes to shift to the (left, right), make cigarettes (more, less) scarce, and cause their relative price to (rise, fall).

14. Import quotas, licensing arrangements, and outright bans on specific goods are forms of government (price, quantity) restrictions.

15. An import quota tends to (lower, raise) price to consumers.

16. If governments put price floors on agricultural goods, a (shortage, surplus) will result.

TRUE-FALSE QUESTIONS

Circle the **T** if the statement is true, the **F** if it is false. Explain to yourself why a statement is false.

T F 1. If supply shifts to the left, given demand, then the equilibrium price and the equilibrium quantity will rise.

T F 2. If demand shifts to the left, given supply, then the equilibrium price and the equilibrium quantity will fall.

T F 3. If both supply and demand shift to the right, then equilibrium price and equilibrium quantity are indeterminate.

T F 4. If the supply of good A increases relative to its demand, then good A is now more scarce, and its relative price will rise.

T F 5. If the published price is constant, but it takes consumers longer to wait in lines, the total price has really risen.

T F 6. If markets are flexible and no market restrictions exist, then surpluses and shortages won't occur, even in the short run.

T F 7. Minimum wage laws are a form of price ceiling.

T F 8. Rent controls help the poor who are looking for apartments, because rents are lower.

T F 9. Black markets, in effect, cause price to rise for certain buyers.

T F 10. Agricultural surpluses arise when governments put price ceilings on such goods.

MULTIPLE CHOICE QUESTIONS

Circle the letter that corresponds to the best answer.

1. Because resources are scarce,
   a. buyers compete with buyers for outputs.
   b. there must be some method for rationing goods.
   c. people cannot have all they want at a zero price.
   d. All of the above

2. If markets are free and prices are flexible,
   a. equilibrium price cannot be established.
   b. shortages and surpluses eventually disappear.
   c. shortages and surpluses can't arise.
   d. equilibrium quantity cannot be established.

3. If demand shifts to the right (given supply), then equilibrium
    a. quantity will rise.
    b. price is indeterminate.
    c. price and equilibrium quantity are indeterminate.
    d. price will fall.

4. If supply shifts to the right (given demand), then equilibrium
    a. quantity will rise.
    b. price will rise.
    c. price and equilibrium quantity will fall.
    d. price and equilibrium quantity rises.

5. If both supply and demand shift to the left, then equilibrium
    a. price is indeterminate and equilibrium quantity rises.
    b. price is indeterminate and equilibrium quantity falls.
    c. price falls and equilibrium quantity falls.
    d. price falls and equilibrium quantity is indeterminate.

6. If the demand for good A falls relative to its supply, then
    a. good A is now relatively more scarce.
    b. good A is now relatively less scarce.
    c. the relative price of good A will rise.
    d. the total price of good A will rise, even if A is not price flexible.

7. If the demand for good B rises relative to its supply, then
    a. good B is now relatively more scarce.
    b. the relative price of good B will rise.
    c. the total price of good B will rise, even if good B is price inflexible.
    d. All of the above

8. If the demand for good A rises relative to its supply, and markets are price flexible, then
    a. no shortage of A can exist in the long run.
    b. no shortage of A can exist in the short run.
    c. the published price of A remains constant, but its total price falls.
    d. the published price of A remains constant, but its total price rises.

9. If the demand for good A rises relative to its supply, and markets are price inflexible, then
    a. a shortage can exist in the short run.
    b. a shortage can exist in the long run.
    c. the published price of A might remain constant, but its total price rises.
    d. All of the above

10. If the demand for economists falls relative to their supply, then
    a. more college students will major in economics.
    b. some economists will change professions.
    c. a shortage of economists will result, in the long run.
    d. All of the above

11. Which of the following can influence how a society rations a specific good?
    a. Price system that rations to the highest bidder.
    b. Political power.
    c. Religion.
    d. All of the above

12. Prolonged shortages arise if
    a. demand increases relative to supply.
    b. price floors are set by governments.
    c. prices are not allowed to rise to equilibrium.
    d. buyers are allowed to compete for goods.

13. Black markets may arise if
    a. price ceilings exist.
    b. price floors exist.
    c. governments do not intervene in the market.
    d. equilibrium price is too low.

14. Rent controls
    a. are a form of price floor.
    b. help the homeless who need apartments.
    c. make tenants less mobile.
    d. reduce litigation in society.

15. If an effective minimum wage is imposed, then
    a. more workers will be unable to find jobs.
    b. the quantity of labor demanded will fall.
    c. some workers will move to sectors not covered by minimum wages.
    d. All of the above

16. Prolonged agricultural surpluses can arise if governments
    a. set price above equilibrium.
    b. institute price floors, or price supports.
    c. purchase the excess supply.
    d. All of the above

MATCHING
Choose the item in Column (2) that best matches an item in Column (1).

	(1)		(2)
a.	price floor	e.	buyer competition
b.	price ceiling	f.	rent control
c.	scarce resources	g.	minimum wage law
d.	nonprice rationing	h.	black market, long lines

WORKING WITH GRAPHS

1. Consider the graphs below, then answer the questions that follow.

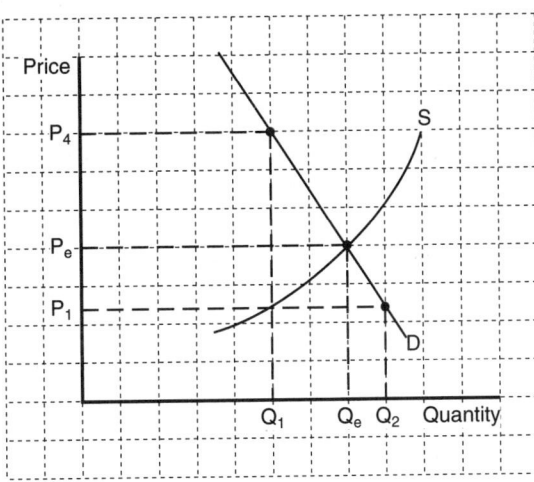

   a. The market clearing price is _____.
   b. If the government imposes a price ceiling at P_1, what will be the quantity supplied? The quantity demanded? What exists at that price?
   c. Given the quantity that will be forthcoming at the permitted price of P_1, what will the actual or black market price be?
   d. Other than via a black market transaction, how can the actual price paid by buyers exceed the permitted price, P_1.
   e. If price had been permitted to rise to equilibrium, what would be the quantity supplied by sellers? Is that amount greater or less than the quantity at P_1? Why?

2. Consider the following supply and demand curves for labor, and then answer the questions.

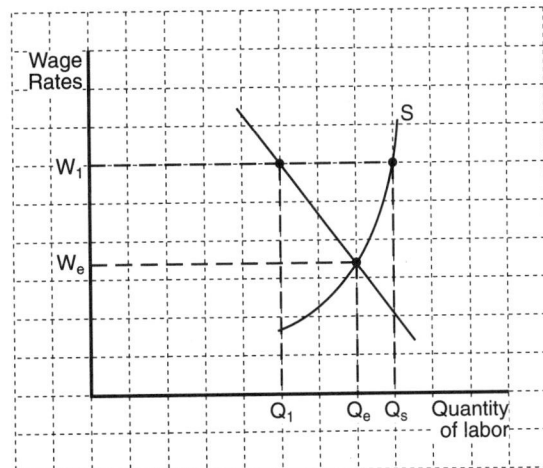

   a. What is the equilibrium wage rate? The equilibrium quantity of labor?
   b. If the government sets a minimum wage rate at W_1, what is the quantity of labor demanded by employers? The quantity of labor supplied by workers? What exists at the minimum wage rate?

c. Is there a shortage or surplus of *jobs*? How might such jobs be allocated (that is, how will employers go about deciding who gets the jobs)?

3. During September of 1989, the then "drug Czar" William T. Bennett, Director of the Office of National Drug Control Policy, and Nobel Prize winning economist Milton Friedman debated the case for the prohibition of drugs in various letters to the editor, in the *Wall Street Journal*. Bennett, who favors prohibition, maintained that if drugs were legalized price would fall, and, therefore, the total amount of drug usage in the U.S. would rise. Friedman, who favors legalization of drugs (to adults) maintained that, once legalized, the demand for drugs would fall because (1) the appeal to people who want the excitement of doing something "taboo" would disappear, and (2) addicts who have only a $2 a day habit have less of an incentive to get other people hooked on drugs (to support their own habit) than if they had a $200 a day habit.

   a. Is Bennett correct when he says that once legalized, price will fall? Why?
   b. If the demand for drugs shifts to the left, and the supply of drugs to the right, what happens to the price of drugs?
   c. If the demand for drugs shifts to the left (Friedman) and the supply of drugs shifts to the right (Bennett), what happens to the equilibrium quantity?
   d. How does your answer to (c) help you decide whether drug usage will rise or fall if drugs are legalized? What information is necessary to resolve this issue?

PROBLEMS

1. In a 1950's movie entitled "Under the Yum-Yum Tree," Jack Lemmon played a lecherous landlord who was extremely wealthy.
   a. How do you imagine he created a shortage of his own luxury apartments, and what criteria do you imagine he used to decide who was able to rent such apartments?
   b. Does rent control force landlords to discriminate in the selection of tenants? What criteria might they use to make such selections?

2. In 1979 the rock group The Who gave a concert in Cincinnati, and 11 people died outside Riverfront Coliseum when the gates were opened and the crowd rushed in to get choice seats. What other method of allocating the scarce resource of choice seats would have prevented this tragic event?

ANSWERS TO CHAPTER 4

COMPLETION QUESTIONS

1. zero; compete
2. fall; fall
3. fall; rise
4. is indeterminate; will rise
5. is indeterminate; will fall
6. more scarce; rise; less scarce; fall
7. risen; risen; time
8. more scarce; rise; increase
9. rationing; highest; nonprice
10. shortages; surpluses
11. ceilings
12. ceiling; reduces; reduces; causes a deterioration in; landlords and low income apartment hunters
13. left; more; rise
14. quantity
15. raise
16. surplus

TRUE-FALSE QUESTIONS

1. F The equilibrium quantity falls.
2. T
3. F Equilibrium quantity rises.
4. F Good A is now less scarce, and its relative price will fall.
5. T
6. F No, surpluses and shortages can exist—in the short run.
7. F They are a price floor.
8. F That group is hurt because they will be discriminated against and because the housing stock diminishes.
9. T
10. F Price ceilings cause shortages.

MULTIPLE CHOICE QUESTIONS

1.d;   2.b;   3.a;   4.a;   5.b;   6.b;   7.d;   8.a;   9.d;   10.b;
11.d; 12.c; 13.a; 14.c; 15.d; 16.d.

MATCHING

a and g; b and f; c and e; d and h

WORKING WITH GRAPHS

1. a. P_e
    b. Q_1; Q_2; shortage
    c. P_4
    d. quality deterioration, long lines that increase opportunity costs
    e. Q_e; greater; a higher price induces sellers to produce more.

2. a. W_e; Q_e
    b. Q_1; Q_s; surplus of labor, or unemployment
    c. shortage; family influence, political power, bribes, racial or gender preference

3. a. He is correct, because the supply curve will shift to the right as the costs and risks of drug-dealing fall.
    b. falls (Therefore Bennett is right.)
    c. It is impossible to predict the net effect on the equilibrium quantity.
    d. The real issue is an empirical one: Will supply rise by more than demand falls, or vice versa?

PROBLEMS

1. a. He set rents far below the market clearing levels, and many people wanted his apartments; he rented only to "lovely young ladies."
    b. Yes; gender, beauty, race, number of children, age, ability to pay, pets, and so on.

2. Instead of "first-come, first-served seating," ticket sellers could have raised the price of choice seats and used assigned seating. Shortly thereafter, the city of Cincinnati passed a resolution that outlawed first-come, first-served seating.

GLOSSARY TO CHAPTER 4

Black market A market in which goods are traded at prices above their legal maximum prices or in which illegal goods are sold.

Import quota A physical supply restriction on imports of a particular good, such as sugar. Foreign exporters are unable to sell in the United States more than the quantity specified in the import quota.

Minimum wage A wage floor, legislated by government, setting the lowest hourly rate that firms may legally pay workers.

Nonprice rationing devices All those methods used to ration scarce goods that are price-controlled. Whenever the price system is not allowed to work, nonprice rationing devices will evolve to ration the affected goods and services.

Price ceiling A legal maximum price that may be charged for a particular good or service.

Price controls Government mandated minimum or maximum prices that may be charged for goods and services.

Price floor A legal minimum price below which a good or service may not be sold. Legal minimum wages are an example.

Price system An economic system in which relative prices are constantly changing to reflect changes in supply and demand for different commodities. The prices of those commodities are signals to everyone within the system about what is relatively scarce and what is relatively abundant.

Rent control The placement of price ceilings on rents in particular cities.

Terms of exchange The terms under which the trading takes place. Usually, the terms of exchange are given by the price at which a good is traded.

Transaction costs All of the costs associated with exchanging, including the informational costs of finding out price and quality, service record, and durability of a product, plus the cost of contracting and enforcing that contract.

Voluntary exchange An act of trading, done on a voluntary basis, in which both parties to the trade are subjectively better off after the exchange.

CHAPTER 5
THE PUBLIC SECTOR AND PUBLIC CHOICE

LEARNING OBJECTIVES

After you have studied this chapter, you should be able to

1. define antitrust legislation, monopoly, spillover or externality, third parties, effluent fee, market failure, property rights, private goods, public goods, principle of rival consumption, exclusion principle, free-rider problem, merit good, demerit good, transfer payment, transfers in kind, marginal and average tax rates, proportional, progressive, and regressive taxation, capital gain, capital loss, retained earnings, and tax incidence;

2. enumerate the five economic functions of government;

3. predict whether a specific good will be overproduced, underproduced, or produced in just the right amount if resources are allocated by the price system;

4. identify which graphs take into account an externality and which do not;

5. list the two ways in which a government can correct for negative externalities;

6. identify the three ways in which a government can correct for positive externalities;

7. list four characteristics of public goods that distinguish them from private goods;

8. distinguish between a marginal and an average tax rate;

9. calculate the tax burden for individuals with different incomes, given different tax structures;

10. identify similarities and differences between market and collective decision making.

CHAPTER OUTLINE

1. The government performs many economic functions that affect the way in which resources are allocated.

a. If a benefit or cost associated with an economic activity spills over to third parties, the price system will misallocate resources; a proper role for government is to correct such externalities.
  i. If a negative externality exists, the price system will overallocate resources to that industry; the government can correct this by taxing or regulating such activities.
  ii. If a positive externality exists, the price system will underallocate resources to that industry; the government can correct this by financing additional production, by providing special subsidies, or by regulation.
b. A legal system that defines and enforces property rights is crucial to the American capitalistic economy.
c. Because a competitive price system transmits correct signals, an important role for government is to promote competition.
d. A price system will underallocate resources to the production of public goods.
  i. Charactcristics of public goods include the following:
     1. They are usually indivisible.
     2. They can be used by more people at no additional cost.
     3. Additional users of public goods do not deprive others of any of the services of the good.
     4. It is difficult to charge individual users a fee based on how much they themselves consume of the public good.
  ii. Because public goods must be consumed collectively, individuals have an incentive to take a free ride and not pay for them.
  iii. Because the price system underproduces public goods, a proper role of government may be to ensure their production.
e. In recent years the government has taken on the economic role of ensuring economy-wide stability: full employment, price stability, and economic growth.

2. The government performs political functions that also affect resource allocation.
  a. Governments subsidize the production of merit goods and tax or prohibit the production of demerit goods.
  b. By combining a progressive tax structure with transfer payments, the government attempts to redistribute income from higher to lower income groups (although many "loopholes" frustrate such a policy).

3. Governments tax in order to obtain revenues to finance expenditures.
  a. The marginal tax rate is the change in the tax payment divided by the change in income.
  b. The average tax rate equals the total tax payment divided by total income.

4. There are three main types of taxation systems.
  a. Under a proportional taxation system, as a person's income rises, the percentage of income paid (rate of taxation) in taxes remains constant.
  b. Under a progressive taxation system, as a person's income rises, the percentage of income paid in taxes rises.
  c. Under a regressive taxation system, as a person's income rises, the percentage of income paid in taxes falls.

5. The federal government imposes income taxes on individuals and corporations, and it collects Social Security taxes and other taxes.
  a. The most important tax in the U.S. economy is the personal income tax; recently some have proposed a consumption tax, which taxes people based on what they actually spend.
  b. The difference between the buying and selling price of an asset, such as a share of stock or a plot of land, is called a capital gain if a profit results, and a capital loss if it doesn't.

6. The corporate income tax is a moderately important source of revenue for the various governments in the U.S. economy.
   a. Corporate stockholders are taxed twice: once on corporate income and again when dividends are received or when the stock is sold.
   b. The incidence of corporate taxes falls on people—consumers, workers, management, and stockholders—not on such inanimate objects as "corporations."

7. An increasing percentage of federal tax receipts is accounted for each year by taxes (other than income) levied on payrolls, such as Social Security taxes and unemployment compensation.

8. Major sources of revenue for states and local governments are sales and excise taxes.

9. The theory of public choice is the study of collective decision making.
   a. Collective decision making involves the actions that voters, politicians, and other interested parties undertake to influence nonmarket choices.
   b. Market and collective decision making are similar in the sense that both involve competition for scarce resources and people motivated by self-interest.
   c. Market and collective decision making are different because the government goods are available for consumption at a price of zero, decisions about what government goods to provide are determined by majority rule, and government can use legally sanctioned force to ensure that its decisions are followed.

KEY TERMS
Antitrust legislation
Monopoly
Capital loss
Capital gain
Merit good
Collective decision making
Incentive structure

Effluent fee
Transfer payment
Retained earnings
Demerit good
Theory of public choice
Government, or political, goods

KEY CONCEPTS
Externality
Third parties
Market failure
Principle of rival consumption
Exclusion principle
Private goods
Public goods

Marginal tax rate
Average tax rate
Transfers in kind
Proportional taxation
Progressive taxation
Regressive taxation

Free-rider problem
Property rights
Public choice
Subsidy
Tax incidence
Tax bracket

COMPLETION QUESTIONS
Fill in the blank, or circle the correct term.

1. The five economic functions of federal government in our capitalistic system are _____, _____, _____, _____, and _____.

2. If there are disputes in an economic arena, the _____ often acts as a "referee" to help settle the dispute.

3. Antitrust legislation, in theory, is supposed to (decrease, promote) competition in the private sector.

4. If externalities are an important result of an economic activity, then the price system is (inefficient, efficient).

5. If Mr. Johnson buys an automobile from General Motors, those people not directly involved in the transaction are considered _____.

6. Pollution is an example of a (negative, positive) externality.

7. When there are spillover costs, a price system will (under, over) allocate resources to the production of the good in question.

8. If third parties benefit from a transaction, then (negative, positive) externalities exist, and the price system will allocate resources (inefficiently, efficiently).

9. Positive and negative externalities are examples of market _____.

10. A government can correct negative externalities by imposing taxes and by _____ the industry or firms in question.

11. A government can correct positive externalities by _____, _____, and _____.

12. If a positive externality exists for good B, a price system will produce too _____ of good B.

13. Public goods have four distinguishing characteristics. They are usually _____; they can be used by more people at _____ additional cost; additional users (do, do not) deprive others of the services of a public good; it is very (easy, difficult) to charge individuals based on how much they used the public good.

14. A free rider has an incentive to (pay, not pay) for a public good.

15. Demerit goods are goods for which society wants to (decrease, increase) production.

16. Many government, or political, goods are provided to consumers at a (zero, positive) price; but the opportunity cost to society of providing government goods is (zero, positive).

17. If the price of an asset rises after its purchase, the owner receives a(n) _____ gain; if the price falls, the owner suffers a(n) _____ loss.

18. The marginal tax rate applies only to the (first, last) tax bracket.

19. The corporate income tax is paid by one or more of the following groups: _____, _____, and _____.

20. In contrast to goods sold in private markets, government goods are not (scarce, explicitly priced).

21. In the government sector, decisions concerning what goods to produce are determined by (majority, proportional) rule.

TRUE-FALSE QUESTIONS

Circle the **T** if the statement is true, the **F** if it is false. Explain to yourself why a statement is false.

T F 1. In the U.S. economy the government plays only a minor role in resource allocation, because the country is capitalistic.

T F 2. Governments provide a legal system, but this important function is not considered an economic function.

T F 3. One aim of antitrust legislation is the promotion of competition.

T F 4. If externalities, or spillovers, exist, then a price system misallocates resources, so that inefficiency exists.

T F 5. If a negative externality exists, buyers and sellers are not faced with the true opportunity costs of their actions.

T F 6. If a positive externality exists when good A is produced, a price system will underallocate resources into the production of good A.

T F 7. One way to help correct for a negative externality is to tax the good in question, because that will cause the price of the good to fall.

T F 8. A price system will tend to overallocate resources to the production of free goods, due to the free-rider problem.

T F 9. Scarcity exists in the market sector, but not in the public sector.

T F 10. If third parties are hurt by the production of good B and they are not compensated, then too many resources have been allocated to industry B.

T F 11. Deciding what is a merit good and what is a demerit good is easily done and does not require value judgments.

T F 12. The federal individual income tax is regressive.

T F 13. The largest source of receipts for the federal government is the individual income tax.

T F 14. In a progressive tax structure, the average tax rate is greater than the marginal tax rate.

T F 15. Positive economics confirms that a progressive taxation system is more equitable than a regressive taxation system.

T F 16. In the United States the tax system that yields the most revenue to all governments combined is the corporate tax.

T F 17. When corporations are taxed, consumers and corporate employees are also affected.

T F 18. Government goods are produced solely in the public sector.

T F 19. A nation's "tax freedom day" is the date when an average resident has earned sufficient income to pay his or her total tax bill for the year.

CHAPTER 5: THE PUBLIC SECTOR AND PUBLIC CHOICE

MULTIPLE CHOICE QUESTIONS
Circle the letter that corresponds to the best answer.

1. Which of the following is **NOT** an economic function of government?
   a. income redistribution
   b. providing a legal system
   c. ensuring economy-wide stability
   d. promoting competition

2. A price system will misallocate resources if
   a. much income inequality exists.
   b. demerit goods are produced.
   c. externalities exist.
   d. All of the above

3. Which of the following does **NOT** belong with the others?
   a. positive externality
   b. negative externality
   c. demerit good
   d. public good

4. The exclusion principle
   a. does not work for public goods.
   b. does not work for private goods.
   c. causes positive externalities.
   d. makes it easy to assess user fees on true public goods.

5. Which of the following statements concerning externalities is true?
   a. If a positive externality exists for good A, A will be overproduced by a price system.
   b. If externalities exist, then resources will be allocated efficiently.
   c. Efficiency may be improved if the government taxes goods for which a positive externality exists.
   d. The output of goods for which a positive externality exists is too low, from society's point of view.

6. Which of the following is **NOT** a characteristic of public goods?
   a. indivisibility
   b. high extra cost to additional users
   c. exclusion principle does not work easily
   d. difficult to determine how each individual benefits from public goods

7. Market failure exists if
   a. Mr. Smith cannot purchase watermelons in his town.
   b. buyers and sellers must pay the true opportunity costs of their actions.
   c. third parties are injured and are not compensated.
   d. the government must provide merit goods.

8. Which of the following will properly correct a negative externality that results from producing good B?
   a. subsidizing the production of good B
   b. letting the price system determine the price and output of good B
   c. forcing buyers and sellers of good B to pay the true opportunity costs of their actions
   d. banning the production of good B

9. Merit and demerit goods
    a. are examples of public goods.
    b. are examples of externalities.
    c. indicate market failure.
    d. are not easily classified.

10. A switch from the current progressive income tax to a national sales tax
    a. would not change our tax system very much.
    b. would lead to more taxes on savings.
    c. would cause the current structure of the IRS to be greatly reduced.
    d. would cause more Internal Revenue agents to be hired.

11. If the government taxes group A and gives to group B, then economic incentives for
    a. group A may be reduced.
    b. group B may be reduced.
    c. both may change so as to reduce output.
    d. All of the above

12. If Mr. Ayres loves good A, he can convey the intensity of his wants if good A is
    a. a private good.
    b. a public good.
    c. not subject to the exclusion principle.
    d. expensive.

13. The free-rider problem exists
    a. for private goods.
    b. for goods that must be consumed collectively.
    c. only if people can be excluded from consumption.
    d. All of the above

14. In a progressive tax structure,
    a. the marginal tax rate exceeds the average tax rate.
    b. equity exists.
    c. the average tax rate rises as income falls.
    d. All of the above

15. Which of the following statements is true?
    a. Under a regressive tax structure, the average tax rate remains constant as income rises.
    b. If upper-income people pay more taxes than lower-income people, equity must exist.
    c. The U.S. federal personal income tax system is progressive.
    d. At very high income levels, the Social Security tax and employee contribution become progressive.

16. The tax incidence of the corporate income tax falls on
    a. corporate stockholders.
    b. corporate employees.
    c. consumers of goods and services produced by corporations.
    d. All of the above

17. Which of the following statements about the Social Security tax is **NOT** true?
    a. It is a progressive tax.
    b. It came into existence in 1935.
    c. It is imposed on employers and employees.
    d. It is a payroll tax.

18. If Mr. Romano faces a 90 percent marginal tax rate,
    a. the next dollar he earns nets him ninety cents.
    b. his total tax payments equal 90 percent of his total income.
    c. he has a strong incentive not to earn extra income.
    d. his average tax rate must be falling.

19. A proportional tax system
    a. is unfair.
    b. cannot be consistent with people's ability to pay such taxes.
    c. means that upper-income people pay smaller percentages of their income in taxes than do lower-income people.
    d. requires upper-income people to pay more tax dollars than lower-income people pay.

20. Which one of the following is true of both market and collective decision making? Within both contexts,
    a. resources are scarce.
    b. people face identical incentive structures.
    c. production and allocation decisions arise from majority rule.
    d. production and allocation decisions arise from proportional rule.

21. Which one of the following is true of government goods?
    a. They are always produced within the public sector.
    b. They are always produced within the private sector.
    c. They are provided free of charge.
    d. They have no opportunity cost.

MATCHING

Choose the item in Column (2) that best matches an item in Column (1).

	(1)		(2)
a.	antitrust legislation	g.	pollution
b.	spillover	h.	externality
c.	positive externality	i.	national defense
d.	negative externality	j.	alcohol
e.	government good	k.	monopoly
f.	demerit good	l.	flu shots

PROBLEMS

1. Complete the following table for three taxes, and then indicate what type of tax each is.

Income	Tax 1		Tax 2		Tax 3	
	Tax paid	Average tax rate	Tax paid	Average tax rate	Tax paid	Average tax rate
$ 1,000	$ 30	_____	$ 10	_____	$ 100	_____
3,000	90	_____	60	_____	270	_____
6,000	180	_____	180	_____	480	_____
10,000	300	_____	400	_____	700	_____
15,000	450	_____	750	_____	900	_____
20,000	600	_____	1200	_____	1000	_____
30,000	900	_____	2100	_____	1200	_____

2. Suppose the above table had a fourth tax as shown below. Find the average and marginal tax rates, and explain what type tax it would be.

Income	Tax paid	Average tax rate	Marginal tax rate
$ 1,000	$ 30	_____	_____
3,000	120	_____	_____
6,000	300	_____	_____
10,000	500	_____	_____
15,000	600	_____	_____
20,000	700	_____	_____
30,000	900	_____	_____

3. One important purely economic function of government is to promote competition, which presumably makes the price system more efficient. During the 1970s, the OPEC oil cartel was able to restrict output dramatically, which permitted the cartel to charge much higher prices and earn higher profits. How did consumers, businesses, and other governments react to the higher relative price of oil? Were such actions rational, from the point of view of the individuals involved? Did such decisions lead to a misallocation of resources from *society's* point of view? (Hint: the OPEC price was artificially high because the cartel reduced output and repressed competition.)

4. In the text, five economic and two political functions of the government were analyzed. Place each of the following governmental activities in one (or more) of these seven categories.

   a. Providing aid to welfare recipients _____
   b. Passing antitrust laws _____
   c. Subsidizing the arts _____
   d. Prohibiting the sale and possession of drugs _____
   e. Providing national defense _____
   f. Enforcing a progressive tax structure _____
   g. Enforcing contracts _____
   h. Providing public education to children _____
   i. Prosecuting fraud _____
   j. Providing funds for AIDS research _____
   k. Creating jobs to reduce unemployment _____

WORKING WITH GRAPHS

1. Consider the graph below, then answer the questions. Assume S represents industry supply and S' includes pollution costs to society as well as industry private costs.

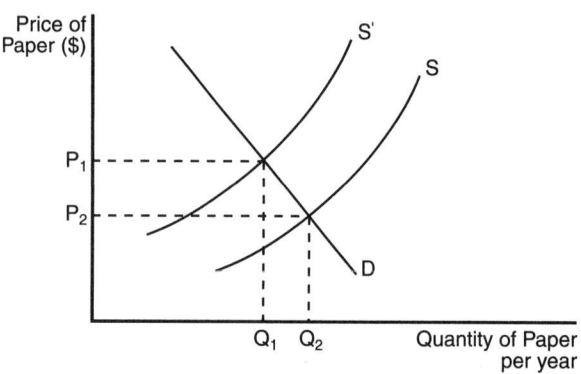

   a. If no governmental intervention takes place, what will be the market equilibrium price? The market equilibrium quantity?

   b. At the market equilibrium quantity (Q_2), which is higher: private costs or social costs?

   c. From *society's* point of view, what is the price that reflects the true opportunity costs of paper? From that same point of view, what is the optimal quantity of paper?

   d. Considering your answers in the above three questions, will a price system produce too little or too much paper?

   e. Does a negative externality or a positive externality exist?

2. Consider the graphs below, then answer the questions that follow. Assume that D represents private market demand and that D' represents benefits that accrue to third parties as well as private benefits.

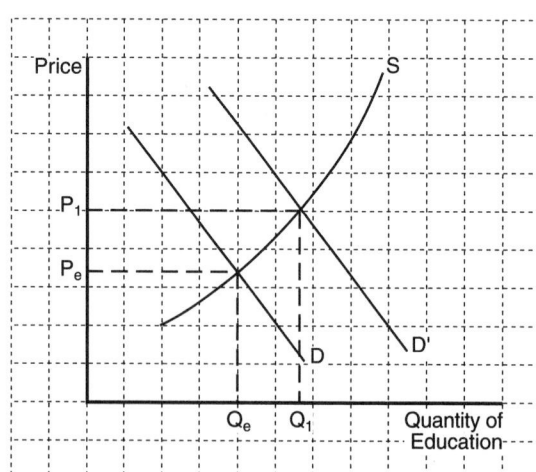

   a. If no government intervention occurs, what will be the market equilibrium price? The market equilibrium quantity?

   b. At the market equilibrium quantity, which is greater, private benefits or social benefits?

c. From *society's* point of view, what is the optimal price and the optimal quantity of education?

d. In this example, does the price system provide too much or too little education?

e. Is there a positive externality or a negative externality for this good?

3. Suppose you know the demand and supply of fertilizer locally, and you have graphed them as shown in the graph that follows. The fertilizer plant that operates in your town is also producing pollution. This pollution is a constant amount per unit of output (proportional to output) at the plant. If the government decides to try to combat the pollution problem by imposing a $20-per-ton tax on fertilizer produced, show graphically what will happen to the fertilizer market. Will the level of pollution in your town be reduced? If so, by how much? If not, can you offer a solution to the pollution problem?

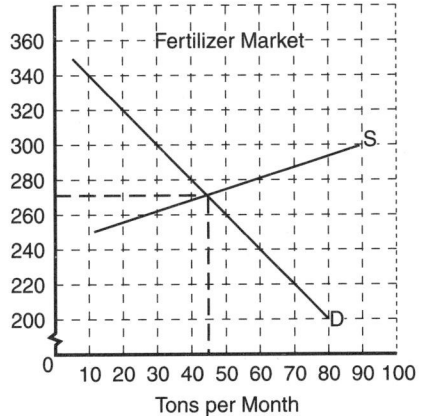

4. a. Draw supply and demand curves for good A, for which no externalities exist, and indicate the optimal quantity of output, the price that reflects the opportunity cost to buyers and sellers of that good, and whether the price system has over-, or under-, or properly allocated resources into industry A.

    b. Draw private supply and demand curves for good B, for which negative externalities exist. Draw another curve on that coordinate system which reflects negative externalities. Indicate the optimal output quantity and price from society's point of view, and compare them to the output quantity and price that would result from a price system.

    c. Draw private supply and demand curves for good C, for which positive externalities exist. Draw another curve which reflects the positive externalities associated with good C. Indicate the optimal price and output quantity of good C from society's point of view, and compare them to the output quantity and price that result from the price system.

ANSWERS TO CHAPTER 5

COMPLETION QUESTIONS

1. providing a legal system, promoting competition, correcting externalities, providing public goods, ensuring economy-wide stability
2. government
3. promote
4. inefficient
5. third parties
6. negative
7. over
8. positive; inefficiently
9. failure
10. regulating
11. subsidizing production, financing production, regulation
12. little
13. indivisible; zero; do not; difficult
14. not pay
15. decrease
16. zero; positive
17. capital; capital
18. last
19. stockholders, consumers, employees
20. explicitly priced
21. majority

TRUE-FALSE QUESTIONS

1. F   Even in capitalist countries the government plays a major role.
2. F   It is an economic function because by enforcing contracts government can promote trade and commerce.
3. T
4. T
5. T
6. T
7. F   A tax will cause the price of the good to *rise,* which is a movement in the correct direction.
8. F   The free-rider problem deals with goods that are *scarce,* but for which the exclusion principle does not work well.
9. F   Scarcity exists in the public sector too; after all, the government uses and allocates scarce goods.
10. T
11. F   Whether or not a good is a merit good requires value judgments.
12. F   It is progressive.
13. T
14. F   For average taxes to rise with income (a progressive tax), the marginal tax rate must exceed the average tax rate.
15. F   "Equitable" requires normative statements.
16. F   No, the personal income tax does so.
17. T
18. F   Governments can also buy privately produced goods for distribution at no charge.
19. T

MULTIPLE CHOICE QUESTIONS

1.a;   2.c;   3.c;   4.a;   5.d;   6.b;   7.c;   8.c;   9.d;   10.c;
11.d; 12.a; 13.b; 14.a; 15.c; 16.d; 17.a; 18.c; 19.d; 21.a;
22. c.

MATCHING

a and k; b and h; c and l; d and g; e and i; f and j

PROBLEMS

1. Tax 1: 3 percent; 3 percent; 3 percent; 3 percent; 3 percent; 3 percent; 3 percent; proportional

   Tax 2: 1 percent; 2 percent; 3 percent; 4 percent; 5 percent; 6 percent; 7 percent; progressive

   Tax 3: 10 percent; 9 percent; 8 percent; 7 percent; 6 percent; 5 percent; 4 percent; regressive

2. ATR: 3 percent; 4 percent; 5 percent; 5 percent; 4 percent; 3.5 percent; 3 percent

   MTR: 3 percent; 4.5 percent; 6 percent; 5 percent; 2 percent; 2 percent; 2 percent

   The average tax rate for this tax initially rises and then falls, as does the marginal tax rate. As a result, this tax is progressive up to an income of $6,000, proportional from there to $10,000, and regressive for levels of income above $10,000. Thus this tax is a combination of all three types of taxes as income varies. Can you graph the ATR and MTR for this tax? Can you think of any taxes that might behave in this manner?

3. Consumers joined car pools, drove less often, bought smaller cars, and endured less comfortable temperatures at home; businesses invested in the production of such oil substitutes as solar energy, nuclear energy, shale oil, coal, etc.; governments subsidized the production of gasohol and shale oil, etc. Such actions were rational because they were responses to a perceived increase in the relative price of oil and its distillates. From society's point of view, such actions led to a misallocation because the lack of competition caused the price system to transmit an incorrect signal. The signal was that oil had become more scarce—but the signal was induced by an artificial restriction of supply.

4. a. redistribution
    b. promoting competition
    c. providing merit goods
    d. discouraging demerit goods
    e. providing public goods
    f. redistribution
    g. providing legal systems
    h. correcting positive externality
    i. providing legal system
    j. correcting negative externality
    k. stabilizing economy

WORKING WITH GRAPHS

1. a. P_2; Q_2
    b. social costs
    c. P_1; Q_1
    d. too much
    e. negative

2. a. P_e; Q_e
   b. social benefits
   c. P_1; Q_1
   d. too little
   e. positive

3. The supply curve after the tax is imposed shifts to S_1—that is, upward by $20 at each quantity. The equilibrium quantity falls from 45 tons per month to below 40 tons per month as a result. Thus the quantity of fertilizer produced has declined by more than 10 percent. This means that the output of pollution has declined by more than 10 percent, because the output of pollution is a constant per unit of output of fertilizer.

   The result of the analysis should not be extended in a general fashion without regard to other possible effects that a tax of this nature might have. We might also wish to consider other factors before imposing a pollution tax. Among these factors are the effects of the increased price of the fertilizer, the likely reduction in employment as a result of the reduced quantity of fertilizer produced, and the ability of alternative methods of pollution control to achieve the same results.

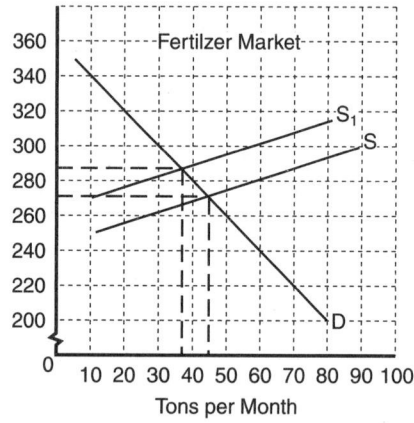

4. a. The market price and the equilibrium quantity are at the socially optimal values because no externalities exist; resources are allocated properly into industry A.
   b. The *new* curve you draw, which reflects a negative externality, should be a supply curve that lies to the left of (above) the original curve, labeled S_1. The optimal price-quantity combination exists where S_1 intersects the demand curve; the socially optimal price is higher than the market price, and the socially optimal quantity is lower than the market quantity.
   c. The new curve should be a demand curve, D_1, that lies to the right of (above) the original demand curve. The socially optimal price-output combination is where D_1 intersects the supply curve; price will be higher and output will be higher than the market price-output combination.

GLOSSARY FOR CHAPTER 5

Antitrust legislation Laws that restrict the formation of monopolies and regulate certain anticompetitive business practices.

Average tax rate The total tax payment divided by total income. It is the proportion of total income paid in taxes.

Capital gain The positive difference between the purchase price and the sale price of an asset. If a share of stock is bought for $5 and then sold for $15, the capital gain is $10.

Capital loss The negative difference between the purchase price and the sale price of an asset.

Collective decision making How voters, politicians, and other interested parties act and how these actions influence nonmarket decisions.

Demerit good A good that has been deemed socially undesirable through the political process. Heroin is an example.

Effluent fee A charge to a polluter that gives the right to discharge into the air or water a certain amount of pollution; also called a *pollution tax*.

Exclusion principle The principle that no one can be excluded from the benefits of a public good, even if that person has not paid for it.

Externality A consequence of an economic activity that spills over to affect third parties. Pollution is an externality.

Free-rider problem A problem that arises when individuals presume that others will pay for public goods, so that, individually, they can escape paying for their portion without a reduction in production.

Government, or political, goods Goods (and services) provided by the public sector; they can be either private or public goods.

Incentive structure The system of rewards and punishments individuals face with respect to their own actions.

Majority rule A collective decision-making system in which group decisions are made on the basis of more than 50 percent of the vote. In other words, whatever more than half of the electorate votes for, the entire electorate has to accept.

Marginal tax rate The change in the tax payment divided by the change in income, or the percentage of additional dollars that must be paid in taxes. The marginal tax rate is applied to the last tax bracket of taxable income reached.

Market failure A situation in which an unrestrained market economy leads to too few or too many resources going to a specific economic activity.

Merit good A good that has been deemed socially desirable through the political process. Museums are an example.

Monopoly A firm that has control over the price of a good. In the extreme case, a monopoly is the only seller of a good or service.

Principle of rival consumption The recognition that individuals are rivals in consuming private goods because one person's consumption reduces the amount available for others to consume.

Private goods Goods that can only be consumed by one individual at a time. Private goods are subject to the principle of rival consumption.

Progressive taxation A tax system in which as income increases, a higher percentage of the additional income is taxed. The marginal tax rate exceeds the average tax rate as income rises.

Property rights The rights of an owner to use and to exchange property.

Proportional rule A decision-making system in which actions are based on the proportion of the "votes" cast and are in proportion to them. In a market system, if 10 percent of the "dollar votes" are cast for blue cars, 10 percent of the output will be blue cars.

Proportional taxation A tax system in which regardless of an individual's income, the tax bill comprises exactly the same proportion.

Public goods Goods for which the principle of rival consumption does not apply; they can be jointly consumed by many individuals simultaneously at no additional cost and with no reduction in quality or quantity.

Regressive taxation A tax system in which as more dollars are earned, the percentage of tax paid on them falls. The marginal tax rate is less than the average tax rate as income rises.

Retained earnings Earnings that a corporation saves, or retains, for investment in other productive activities; earnings that are not distributed to stockholders.

Tax bracket A specified interval of income to which a specific and unique marginal tax is applied.

Tax incidence The distribution of tax burdens among various groups in society.

Theory of public choice The study of collective decision making.

Third parties Parties who are not directly involved in a given activity or transaction.

Transfer payments Money payments made by governments to individuals for which in return no services or goods are concurrently rendered. Examples are welfare, Social Security, and unemployment insurance benefits.

Transfers in kind Payments that are in the form of actual goods and services, such as food stamps, subsidized public housing, and medical care, and for which no goods or services are rendered concurrently.

CHAPTER 6
TAXES, TRANSFERS, AND PUBLIC SPENDING

LEARNING OBJECTIVES

After you have studied this chapter, you should be able to

1. identify the main ways that governments tax sales of goods and services;
2. recognize the difference between static and dynamic tax analysis;
3. explain how levying taxes on goods and services affects market prices and equilibrium quantities;
4. assess the ways in which Medicare affects the incentives to consume medical services;
5. provide reasons why higher government spending on public education has not necessarily improved student performance;
6. discuss the key forces that caused the tremendous rise in Social Security spending;
7. explain why the Social Security Trust Fund is not a stock of savings which the nation may draw upon at a later date, and evaluate how Social Security could be reformed.

CHAPTER OUTLINE

1. The government budget constraint indicates that government spending, transfers, and repayments of borrowed funds are limited to total taxes and user charges the government collects during a given period.

2. When governments attempt to fund their operations by taxing market activities, one issue they must consider is how the tax rates they assess relate to the tax revenues they ultimately receive.
   a. Sales taxes are levied under a system of ad valorem taxation, meaning that the tax is applied to the value of final purchases of a good or service, as determined by its market price, which is the sales tax base. The total sales taxes collected by a government equal the sales tax rate multiplied by the sales tax base, so a sales tax is a proportional tax.
   b. Whereas static tax analysis indicates that a government can unambiguously increase its sales tax collections by boosting the sales tax rate, dynamic tax analysis takes into account the fact that higher tax rates give consumers an incentive to cut back on purchases of goods and services. Dynamic tax analysis indicates that at some point a further increase in the tax

rate reduces the tax base sufficiently to result in lower tax revenues for the government. Consequently, in principle there is a single tax rate at which the government can collect the maximum possible revenues.

3. When contemplating how to structure a system for taxing market transactions, governments must also consider how the taxes they impose affect market prices and equilibrium quantities
   a. Excise taxes are taxes on sales of specific commodities, and governments commonly levy certain excise taxes as a constant tax per unit sold, or a unit tax.
   b. Imposing a unit excise tax on a good or service reduces the net price that a producer receives for each unit sold by exactly the amount of the tax. Following assessment of a unit excise tax, a producer will continue to supply any given quantity only if the price received for that quantity is higher by exactly the amount of the tax. Thus levying a unit excise tax on sales of a good or service causes the market supply curve to shift upward by the amount of the tax.
   c. If the demand curve has its usual downward slope, the upward shift in the supply curve caused by imposing a unit excise tax causes the equilibrium quantity produced and consumed to decline. The market price rises by less than the amount of the tax. Producers pay part of the tax in the form of higher per-unit costs, and consumers pay the remainder of the tax when they purchase the item at the higher market price.

4. Governments use tax revenues to fund expenditures on public goods and merit goods, such as Medicare.
   a. Federal funding of health-care services implies that effective prices that consumers pay for health-care services are less than the prices that health-care providers receive to provide those services, which explains the large quantities of health-care services demanded and supplied under Medicare.
      i. Because the government pays a per-unit subsidy for consuming a health-care service covered by Medicare, the out-of-pocket expense that a Medicare recipient pays for each unit of service—the effective price to the consumer—is relatively low; thus, the quantity of health-care services demanded by Medicare patients is relatively large.
      ii. Suppliers of health-care services are willing to provide the quantity of services demanded by Medicare patients, because the per-unit price they receive is equal to the out-of-pocket expense of Medicare patients plus the government subsidy.
      iii. The Medicare program's total expense for a particular health-care service equals the per-unit subsidy times the quantity of the service demanded by Medicare patients; taxpayers must fund this expense.
   b. In the absence of Medicare subsidies, the equilibrium prices and quantities of health-care services both would be lower than they are with the subsidies provided by this federal program.
      i. This means that Medicare has encouraged increased consumption and production of health-care services.
      ii. As a result, the total expense of the program—the per-unit government subsidy times the quantity of health-care services demanded—is higher than the government estimated using equilibrium quantities as a guide.
   c. To try to contain overall federal spending on Medicare, the government often imposes reimbursement caps, or limits, on specific medical procedures; this can have the unintended effect of worsening patient care and driving the program's costs up even further.

5. Education is another merit good that receives considerable public funding, which currently amounts to more than 5 percent of total U.S. national income.
   a. The basic economics of public funding of education is similar to the economics of public subsidies of health-care programs such as Medicare: Public schools provide educational services at a price below the market price and provide the amount of services demanded at

the below-market price as long as they receive sufficiently large per-unit subsidies from state and local governments.
b. Since 1960, the inflation-adjusted average government subsidy for U.S. primary and secondary education has increased by $5,000 per student, but measures of student performance have remained flat or even declined. A possible explanation is that a higher per-pupil subsidy increases the difference between the per-student cost of providing educational services and the lower valuation of the services by parents and students; thus schools may have allocated resources to activities that have not necessarily enhanced learning.

6. Social Security faces long-term difficulties.
a. People older than 65 already consume more than one-third of the federal government's budget, but the U.S. population continues to age; the population's median age has risen from 28 in 1970 to over 35 today.
b. Social Security's rate of return adjusted for price changes has fallen below rates available on stocks and other savings instruments; thus, many would be better off if they were able to drop out of Social Security and save for retirement on their own.
c. There are several possible reforms that could preserve Social Security as a "social compact" that spans generations of U.S. residents.
 i One of these is to increase Social Security contributions by raising the payroll tax and/or increasing the wage base to which the payroll tax is applied.
 ii. Another reform would entail reducing total benefits, perhaps by means-testing benefit eligibility, reducing benefits to spouses of covered retirees, and/or raising the benefit retirement age.
 iii. Reducing restrictions on immigration by well-trained workers could expand the payroll tax base, thereby increasing current Social Security contributions.
 iv. Using Social Security funds to purchase shares of stock could raise the program's inflation-adjusted rate of return, provided that the returns on stocks remain relatively high; problems with this idea are that returns on stocks are not guaranteed to remain high and that deciding which stocks to buy could pose political difficulties.

KEY TERMS
Ad valorem taxation
Dynamic tax analysis
Excise tax
Government budget constraint

Inflation-adjusted return
Rate of return
Sales taxes
Social Security contributions

Static tax analysis
Tax base
Unit tax

KEY CONCEPTS
Education vouchers
Medicare subsidies

Payroll tax rate
Payroll tax wage base

User charges

COMPLETION QUESTIONS
Fill in the blank, or circle the correct term.

1. Ultimately, all government spending, transfers, and borrowing are primarily financed by (taxes, user charges).

2. The _____ is the limitation that taxes and user charges place on the total amount of government expenditures and transfer payments.

3. Under an *ad valorem* sales tax system, the government applies a (tax rate, constant tax) to the price of an item to determine the tax owed on purchase of the item.

4. _____ tax analysis emphasizes the potential for ever-higher tax rates to induce a reduction in the tax base.

5. (A sales, An excise) tax is levied on purchases of a particular good or service.

6. If the demand and supply curves for an item have their typical shapes, then imposing a unit excise tax on the item results in (a decrease, an increase) in the market price of the item and (a decrease, an increase) in the equilibrium quantity purchased and sold.

7. Because the Medicare program pays a per-unit subsidy for health-care expenses of people covered by the program, the price that they pay for health-care services is (greater than, equal to, less than) the market price, and the quantity of health-care services that they desire to consume is (greater than, equal to, less than) the equilibrium quantity.

8. Because the Medicare program pays a per-unit subsidy for health-care expenses of people covered by the program, the price that providers receive for health-care services is (greater than, equal to, less than) the market price, and the quantity of health-care services they are willing to supply is (greater than, equal to, less than) the equilibrium quantity.

9. An increase in the number of people covered by Medicare will tend to cause the demand for covered health-care services to _____, thereby causing a(n) _____ in both the equilibrium and actual quantities of the service demanded and supplied.

10. When governments provide subsidies to providers of educational services, the result is that the cost of the last unit of services provided is (lower, higher) than the marginal value of the services to parents and students.

11. A government-operated system of educational vouchers provides funds directly to (consumers, producers) of educational services.

12. A rate of return adjusted for the effects of price changes is called the _____ return.

13. The maximum wage earnings subject to the Social Security payroll tax _____ assessed against the earnings is called the wage _____ of the payroll tax system.

14. By increasing the number of workers who make payroll tax contributions, increased (immigration, emigration) could help the long-term prospects of Social Security.

15. Using Social Security contributions to purchase shares of _____ in companies might yield a higher rate of return for Social Security, but there is no guarantee of this.

TRUE-FALSE QUESTIONS
Circle the **T** if the statement is true, the **F** if it is false. Explain to yourself why a statement is false.

T F 1. A sales tax is typically a constant amount charged on the sale of a particular item.

T F 2. Static tax analysis indicates that raising the tax rate by 1 percentage point will always increase tax revenues by an amount equal to 1 percent of the tax base.

T F 3. Every U.S. state government relies on sales taxes to fund at least a portion of its spending and transfer programs.

T F 4. According to dynamic tax analysis, there is likely to be a single tax rate that maximizes government tax collections.

T F 5. Consumers always pay the full amount of a unit excise tax.

T F   6. The price that Medicare patients pay for covered care that they receive is lower than the market price of that care.

T F   7. Not including any administration costs, the direct expense that taxpayers incur in paying the government's share of the total costs of a particular type of care equals the per-unit subsidy that the government pays times the quantity of care demanded under the subsidy.

T F   8. The price that the supplier of a service covered by Medicare receives is higher than the market price of providing that service.

T F   9. If market demand and supply curves have their normal shapes, then the difference between the market price of a health-care service covered by Medicare and the price that Medicare recipients actually pay is equal to the per-unit Medicare subsidy.

T F 10. Since 1960, more than tripling the inflation-adjusted expenditures on U.S. public education has more than doubled the measured performance of U.S. public school students.

T F 11. A key problem for the Social Security system is that the generation following the Baby Boom generation is larger.

T F 12. One possible way to help fund Social Security in future years would be to raise the payroll tax rate for Social Security contributions.

T F 13. One possible way to reduce the future obligations of the Social Security program would be to reduce the initial age that a person is eligible for retirement benefits.

T F 14. Emigration of large numbers of working age people from the United States could do much to reduce the long-term financial problems of Social Security.

T F 15. Loosening restrictions on immigration by poorly trained people who would have trouble finding jobs would do more to aid the long-term financial health of the Social Security program, as compared with allowing immigration by highly qualified workers.

T F 16. There is no guarantee that purchasing stocks with Social Security funds would yield a higher future rate of return for the program.

MULTIPLE CHOICE QUESTIONS
Circle the letter that corresponds to the best answer.

1. If the government establishes a sales tax on a broad set of goods and services by levying a tax rate equal to the market price of each unit purchased, then it uses a system of
   a. unit taxes.
   b. excise taxes.
   c. *ad valorem* taxes.
   d. constant per-unit taxes.

2. The value of goods, services, or incomes subject to taxation is known as the
   a. tax base.
   b. unit base.
   c. *ad valorem* constraint.
   d. government budget constraint.

15. Which one of the following is a common rationale for purchasing shares of stock with Social Security contributions?
    a. In recent years, the inflation-adjusted rate of return has been much higher for stocks than for the Social Security system.
    b. Purchasing shares of stock will permit the federal government to exercise more influence over U.S. companies.
    c. Unlike the rate of return on the current Social Security system, the rate of return on shares of stock is guaranteed.
    d. Buying shares of stock in large quantities will cause the market rate of return on stocks to decline in the future.

MATCHING
Choose an item in Column (2) that best matches an item in Column (1).

(1)

a. Medicare subsidy
b. sales tax
c. *ad valorem* tax
d. dynamic tax analysis
e. static tax analysis

(2)

f. assumes that ever-higher tax rates eventually reduce the tax base
g. assumes that ever-higher tax rates leave the tax base unaffected
h. tax levied as percentage of market price
i. difference between the price a consumer pays and the price received by a seller
j. tax on market prices of many items

PROBLEMS

1. Consider the situation below in a market for health-care services covered by Medicare, where the program pays a per-unit subsidy equal to $20 to suppliers. Suppose that the government expands Medicare and increases the per-unit subsidy by $20. What happens to the market price and equilibrium quantity of health-care services? What happens to the prices paid by Medicare patients and received by health-care providers and to the actual quantity of services consumed and provided under Medicare?

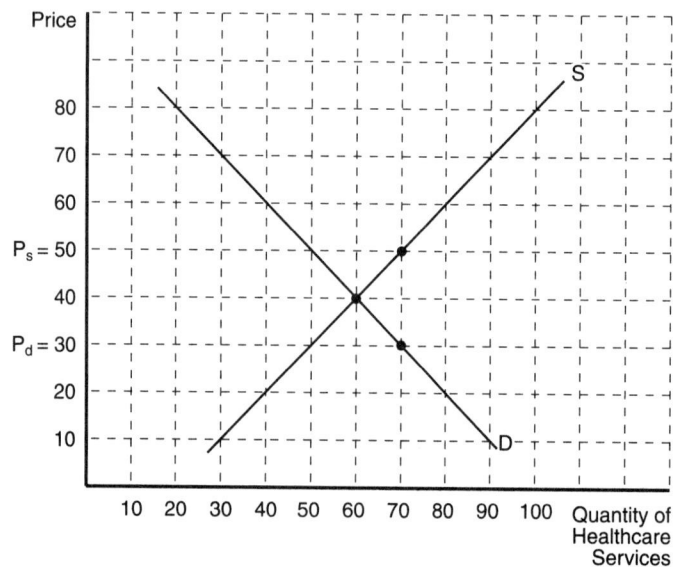

2. In the past, some economists have accused the American Medical Association of restricting the entry of doctors into markets for health-care services. Suppose that these critics are correct and that the AMA successfully reduces the number of doctors offering their services at any given price of health-care services they provide. As a result, the quantity of services provided at each price falls by 20 units. As shown below, the Medicare program pays a per-unit subsidy of $20 to the doctors whom the AMA permits to provide services in this market. Use the diagram below to illustrate the effects of the AMA's action on (1) the market price and equilibrium quantity of health-care services, (2) the price paid by patients covered by Medicare, and (3) the price received by doctors.

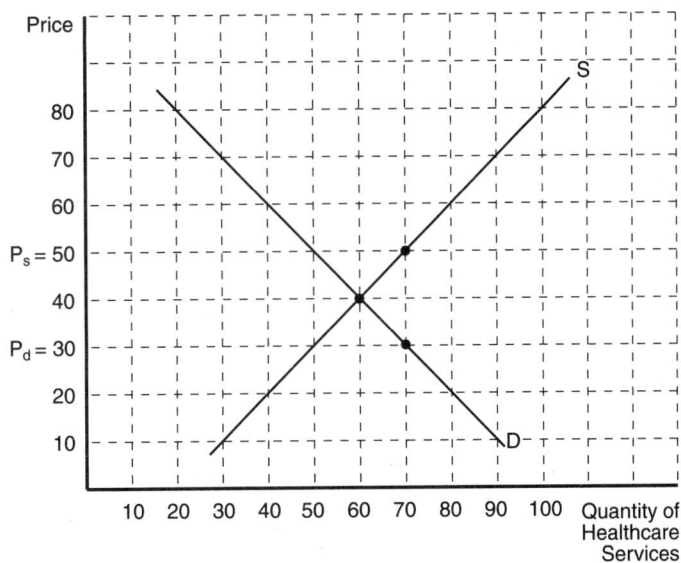

ANSWERS TO CHAPTER 6

COMPLETION QUESTIONS

1. taxes
2. government budget constraint
3. tax rate
4. dynamic
5. An excise
6. an increase; a decrease
7. less than; greater than
8. greater than; greater than
9. increase; increase
10. higher
11. consumers
12. inflation-adjusted
13. rate; base
14. immigration
15. stock

TRUE-FALSE QUESTIONS

1. F   A sales tax covers a large set of items and typically is levied as a fraction of the price.
2. T
3. F   There are no sales taxes in Delaware, Montana, New Hampshire, and Oregon.
4. T
5. F   Consumers pay the full tax only if demand is completely unresponsive to price.
6. T
7. T
8. T
9. F   The Medicare subsidy equals the difference between the price that suppliers receive and the price the recipients pay.
10. F   Measures of student performance have either shown little change or declined.
11. F   The problem is a smaller working age population that must fund benefits for a larger number of older people, including retiring members of the Baby Boom generation.
12. T
13. F   Reducing the retirement age would increase total benefit payments, thereby adding to payouts.
14. F   Emigration of working age residents would reduce the tax base, thereby reducing Social Security contributions.
15. F   This would add to total payouts from the program, which would worsen its problems.
16. T

MULTIPLE CHOICE QUESTIONS

1. c;   2. a;   3. c;   4. a;   5. b;   6. d;   7. b;   8. c;
9. d;   10. d;   11. c;   12. a;   13. b;   14. c;   15. a.

MATCHING

a and i; b and j; c and h; d and f; e and g

PROBLEMS

1. When the subsidy increases, the price paid by Medicare patients declines from $30 per unit to $20 per unit, and the price received by health-care providers increases from $50 per unit to $60 per unit. The quantity of services provided under Medicare increases from 70 units to 80 units. The market price remains equal to $40 per unit, and the equilibrium quantity remains equal to 60 units.

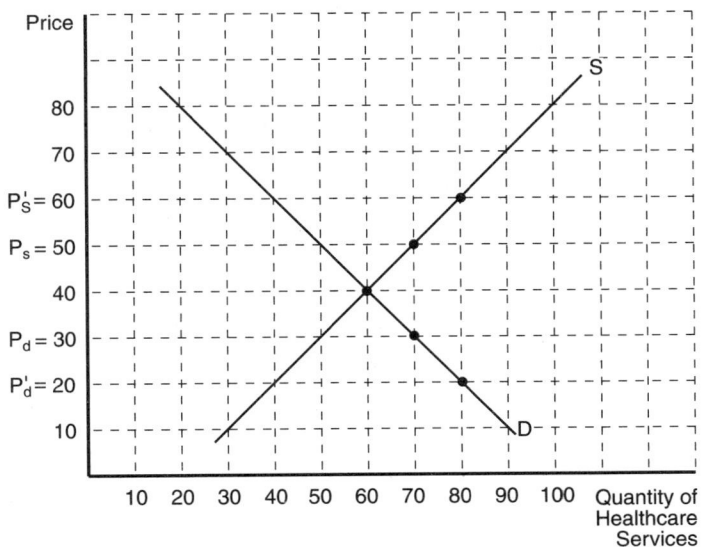

2. The supply curve shifts leftward, so the market price rises from $40 per unit to $50 per unit, and the equilibrium quantity falls from 60 units to 50 units. The subsidy remains equal to $20 per unit, however, so the price paid by Medicare patients increases from $30 per unit to $40 per unit, as does the price received by doctors, from $50 per unit to $60 per unit.

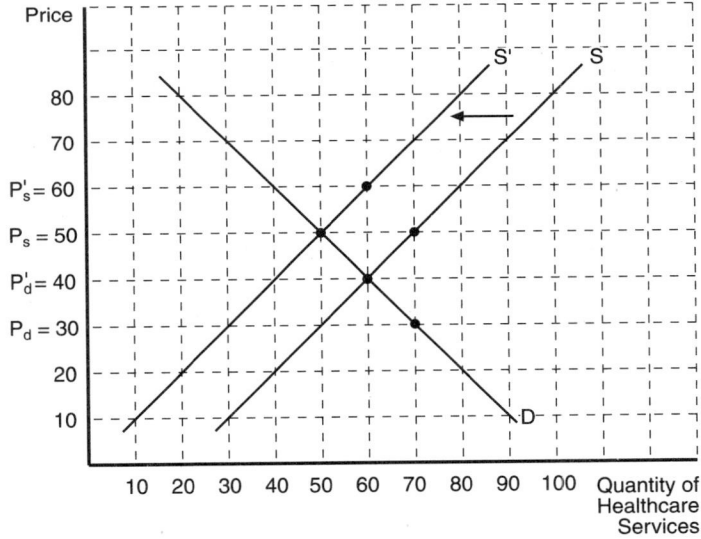

GLOSSARY TO CHAPTER 6

***Ad valorem* taxation:** Assessing taxes by charging a tax rate equal to a fraction of the market price of each unit purchased.

Dynamic tax analysis: Economic evaluation of tax rate changes that recognizes that the tax base eventually declines with ever-higher tax rates, so that tax revenues may eventually decline if the tax rate is raised sufficiently.

Excise tax: A tax levied on purchases of a particular good or service.

Government budget constraint: The limit on government spending and transfers imposed by the fact that every dollar the government spends, transfers, or uses to repay borrowed funds must ultimately be provided by the taxes it collects.

Inflation-adjusted return: A rate of return that is measured in terms of real goods and services; that is, after the effects of inflation have been factored out.

Rate of return: The future financial benefit to making a current investment.

Sales taxes: Taxes assessed on the prices paid on a large set of goods and services.

Social Security contributions: The mandatory taxes paid out of workers' wages and salaries. Although half are supposedly paid by employers, in fact the net wages of employees are lower by the full amount.

Static tax analysis: Economic evaluation of the effects of tax rate changes under the assumption that there is no effect on the tax base, so that there is an unambiguous positive relationship between tax rates and tax revenues.

Tax base: The value of goods, services, incomes, or wealth subject to taxation.

Unit tax: A constant tax assessed on each unit of a good that consumers purchase.

CHAPTER 7
THE MACROECONOMY: UNEMPLOYMENT, INFLATION, AND DEFLATION

LEARNING OBJECTIVES

After you have studied this chapter, you should be able to

1. define the following terms: stock, flow, labor force, unemployment rate, discouraged worker, labor force participation rate, frictional unemployment, cyclical unemployment, seasonal unemployment, structural unemployment, job loser, reentrant, job leaver, new entrant, full employment, inflation, base year, price index, consumer price index, producer price index, GDP deflator, deflation, unanticipated inflation, anticipated inflation, nominal or market rate of interest, real rate of interest, creditor, debtor, cost-of-living adjustments, repricing (menu) cost of inflation, business fluctuations, recession, contraction, depression, expansion, and purchasing power;

2. calculate the unemployment rate, given specific information;

3. recognize the four categories under which an individual is officially counted as being among the unemployed;

4. list the major types of unemployment;

5. distinguish between the effects of anticipated and unanticipated inflation;

6. predict whether a specific group benefits or is economically injured by unanticipated inflation;

7. calculate a price index for a given market basket that includes a small quantity of goods;

8. distinguish between stocks and flows;

9. recognize the types of business fluctuations and list shocks that can cause such fluctuations.

CHAPTER OUTLINE

1. When unemployment exists, the economy is inefficient and households forego goods and services.
   a. The unemployment rate is the percentage of the measured labor force that is unemployed.
   b. The calculation of the official unemployment rate is complicated and somewhat arbitrary.
   c. Four official categories of reasons for unemployment are: job losers, job leavers, reentrants, new entrants.
   d. The unemployment rate varies directly with the average duration of unemployment, other things being constant.
   e. Because of the discouraged worker phenomenon, the official unemployment rate understates true unemployment.

2. Unemployment has been categorized into four types: frictional, cyclical, seasonal, and structural.
   a. Frictional unemployment exists when people are between steady jobs.
   b. Cyclical unemployment is due to recessions and depressions.
   c. Seasonal unemployment results from differences in the demand for labor over the seasons of the year.
   d. Structural unemployment results when resources are reallocated so that individuals with specific skills cannot find jobs for long periods.

3. Employment of every individual in the labor force is generally unattainable because of frictional unemployment.
   a. "Full employment" is said to prevail when only frictional unemployment exists.
   b. The natural rate of unemployment is the unemployment rate that exists in long-run macroeconomic equilibrium, when all workers and employers have fully adjusted to any changes in the economy. Hence, the natural rate of unemployment does not reflect cyclical unemployment.

4. Inflation is a sustained rise in a weighted average of all prices.

5. Measures of inflation include the consumer price index, the producer price index, and the GDP deflator.
   a. The consumer price index measures the cost of an unchanging representative basket of consumer goods through time.
   b. The producer price index measures the cost of an unchanging basket of goods sold in primary markets by producers of commodities in all stages of processing.
   c. The GDP deflator measures the value of all goods and services produced by an economy; the "basket" changes over time.

6. The ill effects of inflation are accounted for mostly by unanticipated inflation; if inflation is fully anticipated, ill effects are slight.

7. The nominal interest rate is (approximately) equal to the sum of the real interest rate and the anticipated inflation rate.
   a. If inflation is unanticipated, creditors are worse off because the real value of their monetary assets falls.
   b. If inflation is unanticipated, debtors are better off because the real value of their monetary debts falls.

8. Besides the transfer of wealth from creditors to debtors during periods of unanticipated inflation, there are other effects.
   a. Inflation reduces the real value of cash holdings.
   b. Resources must be allocated to predict inflation and to avoid its ill effects.

CHAPTER 7: THE MACROECONOMY: UNEMPLOYMENT, INFLATION, AND DEFLATION **73**

9. The ups and downs in economic activity are called business fluctuations.
   a. Inflation tends to be higher during an expansion phase, and unemployment rates are lower.
   b. During a contraction phase unemployment rates are higher and the inflation rate is typically lower.

10. Such external shocks as bad weather and rapid and unanticipated price rises in strategic resources such as oil also cause business fluctuations.

KEY TERMS

Labor force	Unemployment	Recession
Contraction	Full employment	Depression
Job loser	Natural rate of unemployment	Producer price index
Reentrant	Base year	GDP deflator
Job leaver	Price index	Stock
New entrant	Consumer price index	Flow
Expansion		

KEY CONCEPTS

Frictional unemployment	Nominal, or market, rate of interest	Deflation
Discouraged workers		Creditor
Cyclical unemployment	Unanticipated inflation	Debtor
Seasonal unemployment	Cost-of-living adjustments	Purchasing power
Structural unemployment	Business fluctuations	Inflation
Real rate of interest	Anticipated inflation	
Unemployment	Repricing, or menu, cost of inflation	

COMPLETION QUESTIONS
Fill in the blank, or circle the correct term.

1. This chapter deals with unemployment and _____.

2. A(n) _____ is measured per unit of time, while _____ is measured at a given moment in time.

3. The unemployment rate is calculated by dividing the number of unemployed by the sum of the (a) employed plus the (b) _____.

4. If a person is able to work and last looked for a job five weeks ago, she (is, is not) in the labor force.

5. Homemakers are officially considered as (unemployed, employed, not in the labor force).

6. Discouraged workers currently are not looking, but have looked, for a job, and they are counted officially as (employed, unemployed, not in the labor force); because of this official classification, some people believe that the measured unemployment rate (overstates, understates) true unemployment.

7. The percentage of working-age individuals who are in the labor force is called the labor force _____.

8. Unemployment has been categorized into four basic types: frictional, _____, _____, and _____.

9. If imperfect information exists within job markets, then some frictional unemployment is (desirable, undesirable), both from the individual's and society's point of view.

10. _____ unemployment varies with business fluctuations.

11. There is consensus among economists that the current "full" employment rate for the U.S. economy is consistent with about _____ percent actual unemployment.

12. Many of the problems associated with inflation have occurred because the inflation rate was _____.

13. The nominal interest rate equals the real interest rate (plus, minus) the anticipated inflation rate.

14. When inflation is unexpected, (debtors, creditors) benefit at the expense of (debtors, creditors).

15. When unanticipated deflation occurs, debtors are economically (worse, better) off.

16. The personal consumption expenditure index is a price index based on annual surveys of consumer _____.

17. The CPI and the PPI measure the cost of (an unchanging, a changing) basket of goods through time.

18. To the extent that price indexes do not adjust for quality improvements, they (overstate, understate) the true rate of inflation.

19. The Leading Economic Indicators Index includes factors that have been demonstrated to be indicators of (past, future) recessions.

20. Business fluctuations can be caused by _____ shocks such as poor weather or large, unanticipated price rises in key resources such as oil.

TRUE-FALSE QUESTIONS
Circle the **T** if the statement is true, the **F** if it is false. Explain to yourself why a statement is false.

T F 1. Business fluctuations tend to be relatively constant in timing, magnitude, and duration, at least in the United States.

T F 2. The dating of recession and expansion phases is somewhat arbitrary.

T F 3. As inflation occurs, the purchasing power of a unit of money falls.

T F 4. The opportunity costs due to unemployment, in terms of foregone national output, are usually trivial.

T F 5. Homemakers and students are officially counted as part of the labor force.

T F 6. People who are not working and who last looked for a job within the past four weeks are officially unemployed.

T F 7. If the average duration of unemployment rises, other things being constant, the unemployment rate will fall.

T F 8. People not working, who have looked for a job six months ago but are not looking now, are counted as discouraged workers, and therefore are officially unemployed.

T F 9. Reentrants are considered to be unemployed.

T F 10. Because of imperfect information in the labor market, there will always be some frictional unemployment.

T F 11. Anticipated inflation causes fewer economic problems than unanticipated inflation.

T F 12. During periods of correctly anticipated inflation, debtors gain at the expense of creditors.

T F 13. The unemployment rate is higher in the contraction phase of a business cycle.

T F 14. Income is a flow, inflation is a flow, and the number unemployed is a stock.

T F 15. The CPI measures the cost of an unchanging basket of goods and services.

MULTIPLE CHOICE QUESTIONS
Circle the letter that corresponds to the best answer.

1. Business fluctuations are
   a. nonperiodic recurrent fluctuations in overall economic activities.
   b. of similar duration.
   c. of similar magnitude.
   d. All of the above

2. The U.S. labor force includes
   a. the unemployed.
   b. people in mental institutions.
   c. children.
   d. None of the above

3. Which of the following statements is **NOT** a stock concept?
   a. the number of unemployed
   b. national income
   c. the number of job losers
   d. the number of job finders

4. Which of the following is a flow concept?
   a. National income.
   b. Inflation rate.
   c. Consumption.
   d. All of the above

5. Which of the following persons is **NOT** like the others?
   a. job finder
   b. job leaver
   c. new entrant
   d. reentrant

6. Which of the following persons is officially unemployed?
   a. a housewife
   b. a student
   c. a resident in an institution
   d. a nonworking individual who has looked for a job within the past week

7. If the average duration of unemployment rises, other things being constant,
   a. the participation rate will rise.
   b. the unemployment rate will rise.
   c. total employment must fall.
   d. total unemployment must fall.

8. Which of the following statements is false?
   a. It is possible for the total number of employed and the total number of unemployed to rise in the same period.
   b. It is possible for the total number of employed to rise and the unemployment rate to rise in the same period.
   c. If the average duration of unemployment falls, other things being constant, the unemployment rate will fall.
   d. The definitions of employment, unemployment, and labor force are not subject to disagreement among economists.

9. Discouraged workers are officially
   a. unemployed.
   b. employed.
   c. not in the labor force.
   d. in the labor force.

10. If homemakers were counted in the labor force and considered employed, then
    a. the female participation rate would rise.
    b. the overall official unemployment rate would fall.
    c. overall official employment would rise.
    d. All of the above

11. If inflation is anticipated,
    a. it is costless to society.
    b. debtors gain.
    c. it costs less to society than if unanticipated.
    d. creditors gain.

12. Which of the following is **LEAST** like the others?
    a. frictional unemployment
    b. seasonal unemployment
    c. discouraged worker unemployment
    d. cyclical unemployment

13. The teenage unemployment rate is usually high because
    a. many are new entrants in the labor force.
    b. teenagers have a shorter duration of unemployment than adults.
    c. teenagers stay on a given job longer than adults.
    d. All of the above

14. Because there is always frictional unemployment, "full" employment is considered to exist if
    a. every man, woman, and child is working.
    b. everyone age 16 and over is working.
    c. the unemployment rate is relatively small.
    d. the unemployment rate is 12 percent.

15. If the inflation rate is anticipated,
    a. net creditors will be hurt.
    b. inflation may not be a major problem.
    c. net debtors will be hurt.
    d. people will hold more cash than they want to hold.

16. Unanticipated deflation
    a. hurts net debtors.
    b. causes no economic problems.
    c. hurts people who hold cash.
    d. hurts fixed income groups.

17. Which of the following statements is true?
    a. If there is zero anticipated inflation, the nominal interest rate equals the real interest rate.
    b. The real interest rate equals the nominal interest rate plus the anticipated inflation rate.
    c. The real interest rate equals the nominal interest rate divided by the anticipated inflation rate.
    d. Borrowers will not permit lenders to raise the nominal interest rate if all expect that the inflation rate will rise.

18. Which of the following groups is *most* hurt by unanticipated inflation?
    a. workers with cost of living adjustment clauses in their labor contracts
    b. Social Security recipients
    c. workers who sign new work agreements every day
    d. wealthy people who hold much cash in their wall safes

19. The consumer price index
    a. measures the cost of an unchanging basket of goods and services.
    b. does not take into account relative price changes, and therefore is biased.
    c. does not completely account for quality changes, and therefore is biased.
    d. All of the above

MATCHING
Choose the item in Column (2) that best matches an item in Column (1).

(1)	(2)
a. labor force	i. reentrant
b. leading indicators	j. GDP deflator
c. unemployed	k. job gainer
d. price index	l. employed and unemployed
e. employed	m. building permits and vendor deliveries
f. inflation	n. falling purchasing power of money
g. business fluctuation	o. cyclical unemployment
h. recession	p. expansion

PROBLEMS

1. Assume that the employment data (in millions) are

Noninstitutional civilian population (16 and over)	223.190
Resident armed forces	1.568
Civilian labor force	141.600
Unemployment	6.797

   a. Calculate the number of people (16 years old and over) not in the civilian labor force.
   b. Calculate the number of employed civilian workers.
   c. What is the civilian unemployment rate?
   d. In percentage terms (and rounded to the nearest one-hundredth of one percent), what would be the unemployment rate if resident armed forces were counted in the labor force?

2. The CPI measures the cost of an unchanging representative basket of goods and services through time. Suppose the following occur: (a) inflation, (b) an increase in the relative price of energy, and (c) a decrease in the relative price of food. What does the law of demand predict concerning household purchases of food and energy? If households respond predictably, will the CPI overstate or understate the hardships associated with increases in the overall price level?

3. Suppose that an economic slump occurs and that (a) many minorities stop looking for jobs because they know that the probability of finding a job is low, and (b) many people who are laid off start doing such work at home as growing food, painting, repairing their houses and autos, and so on. Which of these events implies that the official unemployment rate overstates unemployment, and which implies the opposite?

4. Consider the following table for an economy that produces only four goods:

Goods & Services	1985 Price	1985 Quantity	2005 Price	2005 Quantity
Pizza	$4	10	$8	12
Cola	12	20	36	15
T-shirts	6	5	10	15
Business equipment	25	10	30	12

   Assuming a 1985 base year,

a. what is nominal GDP for 1985 and for 2005?
b. what is real GDP for 1985? for 2005?
c. what is the implicit GDP price deflator for 1985? for 2005?
d. what is the CPI for 1985? for 2005?

5. From the list below, classify each of the unemployed individuals as representing either (F) frictional, (S) structural, or (C) cyclical unemployment.

   _____ a. James Engine is an auto worker from Detroit who has been laid off because of the recent sharp decline in GDP, which has resulted in a severe decrease in auto sales.
   _____ b. Digs McDuff, from western Kentucky, finds he can no longer get work in the coal mines because of new automated mining techniques.
   _____ c. Priscilla Primm is unable to locate work after finishing her high school education and entering the labor force.
   _____ d. Leroy Cosighn, an aerospace engineer, finds himself unemployed because of large cutbacks in defense spending. Since our space program is also on a tight budget, Leroy hasn't been able to locate alternative work for the past two months.
   _____ e. Oscar Hammerhead, a skilled carpenter, has found himself out of work because of the housing slump brought on by high interest rates and the recession.
   _____ f. Alice Weatherby quits her job as a salesperson out of frustration stemming from her lack of promotion. She begins to look for a management position in a similar work setting.
   _____ g. Patricia Matren reenters the labor force after having a child and is unable to locate suitable work.
   _____ h. Flaps Peterson, an airline pilot, suddenly finds himself laid off because of the dramatic decline in the demand for air transportation caused by the recent recession.

6. Suppose we define our relevant "market basket" of goods as containing the following:

   10 apples   4 pounds of bananas
   7 oranges   2 pineapples

   Suppose we also have the following price information for the years 2000 and 2005:

Fruit	2000	2005
Apples	$ .10 each	$ .18 each
Oranges	.15 each	.23 each
Bananas	.25 per pound	.20 per pound
Pineapples	.50 each	.65 each

   What is the 2005 FPI (fruit price index) using 2000 as the base year? What does this index tell us?

ANSWERS TO CHAPTER 7

COMPLETION QUESTIONS

1. inflation
2. flow; stock
3. unemployed
4. is not
5. not in the labor force
6. not in the labor force; understates
7. participation rate
8. cyclical; seasonal; structural
9. desirable
10. Cyclical
11. 5%
12. unanticipated
13. plus
14. debtors; creditors
15. worse
16. purchases
17. unchanging
18. overstate
19. past
20. external

TRUE-FALSE QUESTIONS

1. F All are highly variable.
2. T
3. T
4. F Costs could be in the hundreds of billions of dollars.
5. F They are not in the offically measured labor force.
6. T
7. F The unemployment rate will rise.
8. F They are, officially, not in the labor force.
9. T
10. T
11. T
12. F Neither group gains or benefits at the expense of the other because the nominal interest rate will reflect the anticipated inflation rate.
13. T
14. T
15. T

MULTIPLE CHOICE QUESTIONS

1.a; 2.a; 3.b; 4.d; 5.a; 6.d; 7.b; 8.d; 9.c; 10.d;
11.c; 12.c; 13.a; 14.c; 15.b; 16.a; 17.a; 18.d; 19.d.

MATCHING

a and l; b and m; c and i; d and j; e and k; f and n; g and p; h and o

PROBLEMS

1. a. 83.158 million   b. 134.803 million   c. 4.80 percent   d. 4.75 percent

2. The law of demand predicts that households will purchase less energy and more food. Because the CPI measures the cost of purchasing an unchanging basket, it will overstate the hardships of inflation.

3. If minorities become discouraged from looking for jobs, they will not be counted as officially unemployed, and therefore the actual employment rate will understate "true" unemployment. If

people perform do-it-yourself activities, they are "really" working, but they won't be counted in the labor force if they quit looking for a job; or they will be counted as unemployed if they continue their job search. Either way, such do-it-yourself activities cause the official unemployment rate to overstate the "true" unemployment rate.

4. a. Nominal GDP for 1985 = ($4) (10) + ($12) (20) + ($6) (5) + ($25) (10) = $560. Nominal GDP for 2005 = ($8) (12) + ($36) (15) + ($10) (15) + ($30) (12) = $1,146.
    b. Real GDP for 1985 = $560. Real GDP for 2005 = ($4) (12) + ($12) (15) + ($6) (15) + ($25) (12) = $618.
    c. GDP deflator for 1985 = (nominal GDP1985/real GDP1985) x 100 = ($560/$560) x 100 = 100.0. GDP deflator for 2005 = (nominal GDP2005/real GDP2005) x 100 = ($1146/$618) x 100 = 185.4.
    d. CPI for 1985 = 100, because it is the base year. CPI for 2005 = (P2005Q1985/P1985Q1985) x 100 = ($850/$310) x 100 = 274.
    P2005Q1985 = ($8) (10) + ($36) (20) + ($10) (5) = $850.
    P1985Q1985 = ($4) (10) + ($12) (20) + ($6) (5) = $310.

5. a. C; b. S; c. F; d. S; e. C; f. F; g. F; h. C

6. | Fruit | Q-05 | P-05 | Total | Q-05 | P-00 | Total |
    |---|---|---|---|---|---|---|
    | Apples | 10 | $.18 | $1.80 | 10 | $.10 | $1.00 |
    | Oranges | 7 | .23 | 1.61 | 7 | .15 | 1.05 |
    | Bananas | 4 | .20 | .80 | 4 | .25 | 1.00 |
    | Pineapples | 2 | .65 | 1.30 | 2 | .50 | 1.00 |
    | | | | $5.51 | | | $4.05 |

Therefore, the FPI = ($5.51 / $4.05) x 100 = 136. This means that on average, fruit is 1.36 times more expensive in 2005 than in 2000. In other words, fruit prices (to the extent that they are measured by our market basket) have risen 36 percent from 2000 to 2005.

Notice that fruit prices have risen at a different rate across different types of fruit. Bananas have actually become cheaper in the above problem. Remember, a price index measures *average* overall tendencies by calculating the ratio of costs of the same "market basket" of goods at *two or more* points in time.

GLOSSARY TO CHAPTER 7

Anticipated inflation The inflation rate that we believe will occur; when it does, we are in a situation of fully anticipated inflation.

Base year The year that is chosen as the point of reference for comparison of prices in other years.

Business fluctuations The ups and downs in overall business activity, as evidenced by changes in national income, employment, and the price level.

Consumer Price Index (CPI) A statistical measure of a weighted average of prices of a specified set of goods and services purchased by wage earners in urban areas.

Contraction A business fluctuation during which the pace of national economic activity is slowing down.

Cost-of-living adjustments (COLAs) Clauses in contracts that allow for increases in specified nominal values to take account of changes in the cost of living.

Cyclical unemployment Unemployment resulting from business recessions that occur when aggregate (total) demand is insufficient to create full employment.

Deflation The situation in which the average of all prices of goods and services in an economy is falling.

Depression An extremely severe recession.

Discouraged workers Individuals who have stopped looking for a job because they are convinced that they will not find a suitable one.

Expansion A business fluctuation in which overall business activity is rising at a more rapid rate than previously or at a more rapid rate than the overall historical trend for the nation.

Flow A quantity measured per unit of time, such as the income you make per week or per year or the number of individuals who are fired every month.

Frictional unemployment Unemployment due to the fact that workers must search for appropriate job offers. This takes time, so they remain temporarily unemployed.

Full employment An arbitrary level of unemployment that corresponds to "normal" friction in the labor market. In 1986, a 6.5 percent rate of unemployment was considered full employment. Today it is assumed to be 5 percent or possibly even less.

GDP deflator A price index measuring the changes in prices of all new goods and services produced in the economy.

Inflation The situation in which the average of all prices of goods and services in an economy is rising.

Job leaver An individual in the labor force who quits voluntarily.

Job loser An individual in the labor force whose employment was involuntarily terminated.

Labor force Individuals aged 16 years or older who either have jobs or are looking and available for jobs; the number of employed plus the number of unemployed.

Labor force participation rate The percentage of noninstitutional working-age individuals who are employed or seeking employment.

Leading indicators Events that have been found to exhibit changes before changes in business activity.

Natural rate of unemployment The rate of unemployment that is estimated to prevail in long-run macroeconomic equilibrium, when all workers and employers have adjusted to any changes in the economy.

New entrant An individual who has never held a full-time job lasting two weeks or longer but is now seeking employment.

Nominal rate of interest The market rate of interest expressed in today's dollars.

Price index The cost of today's market basket of goods expressed as a percentage of the cost of the same market basket during a base year.

Producer Price Index (PPI) A statistical measure of a weighted average of prices of goods and services that firms produce and sell.

Purchasing power The value of money for buying goods and services. If your money income stays the same but the price of one good that you are buying goes up, your effective purchasing power falls, and vice versa.

Real rate of interest The nominal rate of interest minus the anticipated rate of inflation.

Recession A period of time during which the rate of growth of business activity is consistently less than its long-term trend or is negative.

Reentrant An individual who used to work full time but left the labor force and has now reentered it looking for a job.

Repricing, or menu, cost of inflation The cost associated with recalculating prices and printing new prices when there is inflation.

Seasonal unemployment Unemployment resulting from the seasonal pattern of work in specific industries, usually due to seasonality in demand or to changing weather conditions, rendering certain work difficult, if not impossible, as in the agriculture, construction, and tourist industries.

Stock The quantity of something, measured at a given point in time—for example, an inventory of goods or a bank account. Stocks are defined independently of time, although they are assessed at a point in time.

Structural unemployment Unemployment resulting from a poor match of workers' abilities and skills with current requirements of employers.

Unanticipated inflation Inflation at a rate that comes as a surprise, either higher or lower than the rate anticipated.

Unemployment The total number of adults (aged 16 years and older) who are willing and able to work and who are actively looking for work but have not found a job.

CHAPTER 8
MEASURING THE ECONOMY'S PERFORMANCE

LEARNING OBJECTIVES

After you have studied this chapter, you should be able to

1. define total income, final goods and services, gross domestic product, intermediate goods, value added, expenditure approach, income approach, durable consumer goods, nondurable consumer goods, gross private domestic investment, producer durables or capital goods, fixed investment, inventory investment, depreciation or capital consumption allowance, net domestic product, net investment, indirect business taxes, nonincome expense items, national income, personal income, disposable personal income, nominal values, real values, constant dollars, purchasing power parity, and foreign exchange rate;

2. distinguish between flows and stocks, intermediate and final goods, durable and nondurable goods, nominal and real values, goods and services, and gross and net private domestic investment;

3. recognize whether a transaction is or is not included in gross domestic product;

4. recognize whether the inability to include an activity in gross domestic product causes our measure to overstate or understate output and/or economic welfare;

5. list the three general categories of purely financial transactions;

6. distinguish between the expenditure approach and the income approach to deriving gross domestic product;

7. recognize the major components of GDP, using the expenditure approach and using the income approach;

8. derive GDP, NDP, NI, PI, and DPI when given sufficient information;

9. convert nominal GDP into real GDP, given the GDP price deflator;

10. point out problems with measuring GDP and with making international comparisons of GDP;

11. recognize some determinants of the size of a nation's underground economy.

CHAPTER OUTLINE

1. National income accounting is a measurement system used to estimate national income and its components.

2. In order to eliminate the effects of inflation and deflation, statisticians convert nominal GDP into real GDP by dividing the former by a price index.

3. Gross domestic product (GDP) is the market value of all the final goods and services produced by factors of production located within a nation's borders.
   a. In order to avoid double counting, only final goods and services are counted in GDP determination.
   b. Because nonproductive transactions do not contribute to output or to economic welfare, they are excluded from GDP determination.
      i. Such financial transactions as (a) purchases and sales of securities and (b) private and public transfers are nonproductive activities; they are therefore excluded from GDP determination.
      ii. The transfer of used goods is considered a nonproductive activity because by definition used goods are produced (and counted) in a previous period.
      iii. Other transactions excluded from GDP determination are homemaker activities, underground activities, most illegal activities, and do-it-yourself activities; in principle most of these activities *should* be counted in GDP determination, but they are difficult to measure.

4. There are two basic approaches to measuring GDP: the expenditure approach and the income approach.

5. The expenditure approach measures GDP by summing the value of household consumption expenditures, gross private domestic investment, government expenditures, and net exports.
   a. Household consumption expenditures fall into three categories: durable goods, nondurable goods, and services.
   b. Gross private domestic investment equals the sum of fixed investment, inventory investment, and consumer expenditures on new residential structures.
   c. Government expenditures equal the cost of goods and services purchased by governments, because such goods are usually provided at a zero price to users.
   d. Net exports equal the value of exports minus the value of imports.
   e. Net domestic product (NDP) equals gross domestic product minus depreciation.
   f. Indirect business taxes include the value of excise, sales, and property taxes.
   g. Depreciation is also referred to as capital consumption allowance; indirect business taxes plus depreciation equal nonincome expense items.

6. Other components of national income accounting are national income (NI), personal income (PI), and disposable personal income (DPI).
   a. National income equals NDP minus indirect business taxes; using the income approach, NI equals the sum of all factor payments to resource owners.
   b. Personal income equals NI *minus* corporate taxes, Social Security contributions, and undistributed corporate profits, *plus* public transfer payments; using the income approach, PI equals the amount of income that households actually receive before they pay their personal income taxes.
   c. Disposable personal income equals PI minus personal income taxes and nontax payments; DPI equals the income that households have to spend for consumption and saving.

7. Because we are really interested in variations in the real output of the economy, nominal GDP is divided by a GDP price deflator in order to obtain real GDP.

8. Real GDP divided by population yields per capita real GDP; this latter statistic provides a better measure of a nation's living standard.

9. The official GDP measure *underestimates* national output and economic welfare because it does not take into account do-it-yourself activities, homemakers' services, some illegal activities, and underground economy activities. In general, an underground economy will be more important the higher are marginal tax rates on income and the higher are legally mandated benefits that employers must pay to workers.

10. Because GDP is difficult to measure, international GDP comparisons are very difficult; a recent improvement is the purchasing power parity concept, which takes into account the costs of goods and services that are not traded internationally—and therefore are not reflected in foreign exchange rates.

KEY TERMS
Total income
Durable consumer good
Nondurable consumer good
Net investment
Gross domestic product
Gross private domestic
    investment
Fixed investment
Inventory investment
Indirect business taxes
Personal income
Disposable personal income
Net domestic product

KEY CONCEPTS
Intermediate goods
Expenditure approach
Income approach
Producer durables
Nominal values
Depreciation
Nonincome expense items
Constant dollars
Value added
Real values
Purchasing power parity
Foreign exchange rate
Capital consumption allowance
Final goods and services

COMPLETION QUESTIONS
Fill in the blank, or circle the correct term.

1. Inflation causes us to (understate, overstate) the value of output and economic welfare, while deflation causes us to (understate, overstate) such values. For that reason, economists attempt to correct for price level changes; they attempt to convert the less accurate (nominal, real) values into the more accurate (nominal, real) values.

2. GDP represents the total market value of all (final, final and intermediate) goods and services produced during a year.

3. Values that are calculated at a moment in time are referred to as _____; values that are calculated over a time interval are _____ values. Examples of flows include income, consumption, and saving; examples of stocks include _____ and _____.

4. In order to avoid double counting, _____ goods are not counted; only final goods are included in national income accounting.

5. Nonproductive activities, such as financial transactions, (are, are not) counted in GDP determination; if you sell your 4-year-old car to your friend, this activity (is, is not) counted in GDP determination.

6. When a person receives a Social Security payment, the value (is, is not) counted as a productive activity; counting transfers as a productive activity would be an example of _____ counting.

7. Do-it-yourself activities, homemakers' activities, and (legal) underground economy activities (are, are not) productive activities; such activities (are, are not) counted in the official GDP figures; for that reason the GDP figures (overestimate, underestimate) national output and economic welfare.

8. The two basic methods of GDP determination are the _____ approach and the _____ approach. The expenditure approach to national income includes the sum of the values of _____, _____, _____, and _____. The income approach estimates national income by summing _____, _____, _____, and _____.

9. Because many government goods and services are provided to users free of charge, such items are valued at their _____ of production.

10. Consumer durable goods are arbitrarily defined as items that last more than _____ year(s).

11. Net investment equals gross private domestic investment minus _____.

12. Net exports are identical to total exports (plus, minus) total imports.

13. Gross domestic product minus depreciation equals _____; nonincome expense items include indirect business taxes and _____; national income equals NDP minus _____; personal income minus personal income taxes and nontax payments equals _____.

14. When nominal GDP is divided by the _____, real GDP is determined; when real GDP is divided by population, _____ is determined.

15. GDP accounting has been criticized. GDP understates productive activities and economic welfare because it (includes, excludes) household production, do-it-yourself activities, and otherwise legal activities in the _____ economy. Because various things such as pollution (are, are not) subtracted from GDP, GDP overstates economic welfare.

16. The purchasing power parity approach to making international comparisons of living standards (does, does not) consider relative costs of goods that are not traded internationally.

TRUE-FALSE QUESTIONS
Circle the **T** if the statement is true, the **F** if it is false. Explain to yourself why a statement is false.

T F 1. Inflation causes us to overstate national income and output.

T F 2. Gross domestic product is a stock concept.

T F 3. Both final and intermediate goods are counted when measuring GDP.

T F 4. Homemakers' activities are nonproductive transactions.

T F 5. When Mr. Smith purchases a share of stock, investment rises; therefore GDP rises.

T F 6. Public transfers are counted in GDP, but private transfers are not.

T F 7. A nation's underground economy becomes larger as marginal tax rates rise on income.

T F 8. Whether or not a good is durable is an arbitrary decision.

T F 9. In the expenditure approach, the value G equals the sum of all the receipts governments realize from the sale of their services, plus taxes.

T F 10. When a person purchases a new pair of socks, consumption takes place in the official GDP accounts.

T F 11. If net exports rise, other things being constant, then GDP rises.

T F 12. Corporate income taxes are a form of indirect business tax.

T F 13. The sum of household consumption plus household saving equals disposable personal income.

T F 14. GDP minus depreciation equals net private domestic investment.

T F 15. The dollar value of total output computed using the expenditure approach is identical to the dollar value of the total income measured by the income approach because of the way in which profit is defined.

MULTIPLE CHOICE QUESTIONS
Circle the letter that corresponds to the best answer.

1. Concerning real vs. nominal values,
    a. people respond to changes in real values.
    b. economists attempt to convert real into nominal values.
    c. current values are real values.
    d. nominal values have been adjusted for changes in the price level.

2. Gross domestic product includes
    a. only intermediate goods and services.
    b. only final goods and services.
    c. both intermediate and final goods.
    d. neither intermediate nor final goods.

3. In order to avoid overstating national output and income,
    a. intermediate goods are ignored.
    b. used good transactions between nonbusinesses are ignored.
    c. public and private transfers are ignored.
    d. All of the above

4. Which of the following is a nonproductive transaction?
   a. Mr. Gauss gives his niece $50 for her birthday.
   b. Mrs. Patullo cooks for her family.
   c. Mrs. Arianas is a waitress in the "underground" economy.
   d. All of the above

5. Which of the following activities is ignored in the official national income accounts?
   a. Mr. Pulsinelli gives his son $500 for Christmas.
   b. Mrs. Pulsinelli sells her used car to the Harrymans.
   c. Beth Pulsinelli paints her own house.
   d. All of the above

6. *Analogy*: Consumption is to flow as _____ is to stock.
   a. inventory value
   b. GDP
   c. NDP
   d. saving

7. When Capra purchases a bottle of French wine,
   a. consumption falls.
   b. investment rises by the purchase price.
   c. consumption rises by the purchase price.
   d. net exports fall.

8. Net investment equals
   a. GDP minus capital consumption allowance.
   b. gross private domestic investment plus depreciation.
   c. gross private domestic investment minus depreciation.
   d. planned saving minus net saving.

9. If total exports exceed total imports, other things being constant, then
   a. total expenditures fall.
   b. net exports are positive.
   c. GDP falls.
   d. investment rises.

10. GDP minus depreciation equals
    a. net investment.
    b. capital consumption allowances.
    c. NDP.
    d. NI.

11. Which is **NOT** a component of indirect business taxes?
    a. sales taxes
    b. excise taxes
    c. corporate income taxes
    d. property taxes incurred by business persons

12. Which of the following is a nonincome expense item?
    a. depreciation
    b. excise and sales taxes
    c. property taxes incurred by businesspersons
    d. All are nonincome expense items.

13. National income
    a. minus depreciation equals NDP.
    b. plus depreciation plus indirect business taxes equals GDP.
    c. minus inflation equals real GDP.
    d. plus transfer payments equals PI.

14. Which of the following is **NOT** included in national income?
    a. corporate taxes
    b. Social Security taxes
    c. transfer payments
    d. undistributed corporate profits

15. Which of the following is a transfer payment?
    a. Mr. Farano pays his son for painting the house.
    b. Mr. Scheifele gets paid for tending bar but does not declare his income.
    c. Mrs. Niemeck gets paid by a state government for teaching.
    d. Mrs. Carson receives Social Security benefits.

16. Which of the following best represents a nation's standard of living?
    a. nominal GDP
    b. real GDP
    c. per capita real GDP
    d. per capita nominal GDP

17. Which of the following activities is **NOT** considered in GDP determination and therefore causes economic welfare to be overestimated?
    a. do-it-yourself activities
    b. pollution damage
    c. homemaker activities
    d. private and public transfers

18. Forecasters have difficulty in predicting economic recessions because
    a. data from government agencies are often revised.
    b. government price indices are unreliable.
    c. it is difficult to measure recent technological improvements.
    d. All of the above

MATCHING
Choose the item in Column (2) that best matches an item in Column (1).

	(1)		(2)
a.	expenditure approach	j.	depreciation
b.	income approach	k.	NDP
c.	intermediate good	l.	C + I + G + net exports
d.	capital consumption allowance	m.	wages + rents + profits + interest payments
e.	GDP minus depreciation	n.	flour used by a baker
f.	nonincome expense item	o.	bread used by a family
g.	final good	p.	price-level adjusted GDP
h.	constant dollars	q.	capital good
i.	producer durable	r.	indirect business taxes

PROBLEMS

1. What happens to the official measure of GDP if
   a. a woman marries her butler?
   b. an addict marries his cocaine supplier?
   c. homemakers perform the same jobs but switch houses and charge each other for their services?

2. What happens to economic welfare in the three examples in problem 1 above?

3. a. In the table below, calculate real GDP for each of the years indicated.

Year	Nominal GDP	GDP Deflator	Real GDP (1996 dollars)
2001	10,100.1	109.3	_____
2002	10,625.5	111.0	_____
2003	11,224.8	113.2	_____
2004	11,925.2	114.3	_____
2005	12,826.0	115.8	_____

   b. Interpret what a GDP deflator of 115.8 for the year 2005 means.
   c. Determine whether or not inflation occurred over the 2001-2005 period.

4. Suppose you are given the following information about some hypothetical economy and its national income accounts. Use this information to answer the questions that follow. (Amounts are in billions of dollars.)

Indirect business taxes	$ 148
Corporate profits	101
Corporate income taxes	56
Retained earnings	24
Proprietors' income	73
Rents and interest earned (R + I)	98
Exports	18
Imports	10
Net domestic product	1,436
Government expenditures on goods & services	323
Transfer payments	230
Social Security contributions	120
Consumption expenditures	1,055
Gross investment	220
Disposable personal income	1,123

   a. Find GDP.
   b. Find depreciation (capital consumption allowance).
   c. Find domestic factor income receipts.
   d. Find wages and salaries.
   e. Find personal income.
   f. Find personal income taxes.
   g. Find net exports.

5. Suppose you own a small skateboard factory that has sales, expenses, and profits as shown below.

Total sales	$25,000
Expenses	
Wages and salaries	9,000
Interest on loans	800
Rent	3,200
Raw materials	7,000
Tools and equipment	1,000
Profits	$ 4,000

What is the value added to GDP of the productive activities of your firm?

ANSWERS TO CHAPTER 8

COMPLETION QUESTIONS

1. overstate; understate; nominal; real
2. final
3. stocks; flow; inventory value, bank accounts
4. intermediate
5. are not; is not
6. is not; double
7. are; are not; underestimate
8. expenditure; income; consumption, investment, government expenditures, net exports; wages, interest, rents, profits
9. cost
10. three
11. depreciation (or capital consumption allowance)
12. minus
13. NDP; depreciation; indirect business taxes; disposable personal income
14. GDP deflator; per capita real GDP
15. excludes; underground; are not
16. does

TRUE-FALSE QUESTIONS

1. T
2. F   GDP is measured per unit of time; hence it is a flow concept.
3. F   Only final goods are counted, to avoid double counting.
4. F   They are productive; if someone outside the family did them, you would probably have to pay for such services.
5. F   Common stock purchases are simply financial transactions.
6. F   Neither is counted, as should be the case.
7. T
8. T
9. F   G equals the value (at cost) of government purchases of goods and services.
10. T
11. T
12. F   They are direct taxes.
13. T
14. F   It equals NDP.
15. T

MULTIPLE CHOICE QUESTIONS

1.a;   2.b;   3.d;   4.a;   5.d;   6.a;   7.d;   8.c;   9.b;   10.c;
11.c; 12.d; 13.b; 14.c; 15.d; 16.c; 17.b; 18.d.

MATCHING

a and l; b and m; c and n; d and j; e and k; f and r; g and o; h and p; i and q

PROBLEMS

1. a. falls
   b. unaffected, because illegal activities not counted anyway
   c. rises

2. a. remains constant
   b. difficult to tell; it depends on one's value judgment
   c. remains constant

3. a.
Year	Real GDP (1996 dollars)
2001	9,240.7
2002	9,572.5
2003	9,915.9
2004	10,433.2
2005	11,076.0

   b. The overall price level in 2005 was about 15.8 percent higher than it was in 1996 (the base year).
   c. Inflation occurred in every year over that period because the GDP deflator went up every year.

4. a. To calculate GDP: C + I + G + X = 1,055 + 220 + 323 + 8 = 1,606
   b. To calculate depreciation: GDP - NDP = 1,606 – 1,436 = 170
   c. To calculate domestic factor income: NDP – indirect business taxes = 1,436 - 148 = 1,288
   d. To calculate wages and salaries: domestic factor income - corp. profits - prop. income - R & I = 1,288 - 101 - 73 - 98 = 1,016
   e. To calculate PI: domestic factor income - corp. tax - soc. sec. - ret. earn. + trans. pay. = 1,288 - 56 - 120 - 24 + 230 = 1,318
   f. To calculate personal income taxes: PI - DPI = 1,318 – 1,123 = 195
   g. To calculate net exports: X = exports - imports = 18 - 10 = 8

5. Value added is measured by the difference in the value of the intermediate goods (in this case raw materials) used to produce a product and the final value of that product. Thus,

   value added = total sales - cost of raw materials, or
   value added = $25,000 - $7,000 = $18,000.

GLOSSARY TO CHAPTER 8

Capital consumption allowance Another name for depreciation, the amount that businesses would have to save in order to take care of the deterioration of machines and other equipment.

Constant dollars Dollars expressed in terms of real purchasing power using a particular year as the base or standard of comparison, in contrast to current dollars.

Depreciation Reduction in the value of capital goods over a one-year period due to physical wear and tear and also to obsolescence; also called *capital consumption allowance.*

Disposable personal income (DPI) Personal income after personal income taxes have been paid.

Durable consumer goods Consumer goods that have a life span of more than three years.

Expenditure approach Computing national income by adding up the dollar value at current market prices of all final goods and services.

Final goods and services Goods and services that are at their final stage of production and will not be transformed into yet other goods or services. For example, wheat is not ordinarily considered a final good because it is used to make a final good, bread.

Fixed investment Purchases by businesses of newly produced producer durables, or capital goods, such as production machinery and office equipment.

Foreign exchange rate The price of one currency in terms of another.

Gross domestic income (GDI) The sum of all income—wages, interest, rent, and profits—paid to the four factors of production.

Gross domestic product (GDP) The total market value of all final goods and services produced by factors of production located within a nation's borders.

Gross private domestic investment The creation of capital goods, such as factories and machines, that can yield production and hence consumption in the future. Also included in this definition are changes in business inventories and repairs made to machines or buildings.

Income approach Measuring national income by adding up all components of national income, including wages, interest, rent, and profits.

Indirect business taxes All business taxes except the tax on corporate profits. Indirect business taxes include sales and business property taxes.

Intermediate goods Goods used up entirely in the production of final goods.

Inventory investment Changes in the stocks of finished goods and goods in process, as well as changes in the raw materials that businesses keep on hand. Whenever inventories are decreasing, inventory investment is negative; whenever they are increasing, inventory investment is positive.

Investment Any use of today's resources to expand tomorrow's production or consumption.

National income (NI) The total of all factor payments to resource owners. It can be obtained by subtracting indirect business taxes from NDP.

National income accounting A measurement system used to estimate national income and its components; one approach to measuring an economy's aggregate performance.

Net domestic product (NDP) GDP minus depreciation.

Net investment Gross private domestic investment minus an estimate of the wear and tear on the existing capital stock. Net investment therefore measures the change in our capital stock over a one-year period.

Nominal values The values of variables such as GDP and investment expressed in current dollars, also called *money values*; measurement in terms of actual market prices at which goods and services are sold.

Nondurable consumer goods Goods used by consumers that are used up within three years.

Nonincome expense items The total of indirect business taxes and depreciation.

Personal income (PI) The amount of income that households actually receive before they pay personal income taxes.

Producer durables, or capital goods Durable goods having an expected service life of more than three years that are used by businesses to produce other goods and services.

Purchasing power parity Adjustments in exchange rate conversions that take into account the differences in the true cost of living across countries.

Real values Measurement of economic values after adjustments have been made for changes in the average of prices between years.

Services Mental or physical labor or help purchased by consumers. Examples are the assistance of physicians, lawyers, dentists, repair personnel, housecleaners, educators, retailers, and wholesalers; things purchased or used by consumers that do not have physical characteristics.

Total income The yearly amount earned by the nation's resources (factors of production). Total income therefore includes wages, rent, interest payments, and profits that are received, respectively, by workers, landowners, capital owners, and entrepreneurs.

Value added The dollar value of an industry's sales minus the value of intermediate goods (for example, raw materials and parts) used in production.

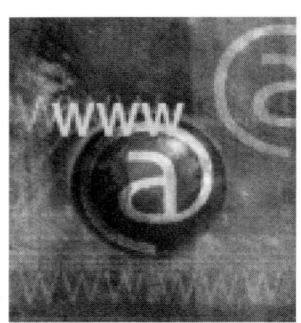

CHAPTER 9
GLOBAL ECONOMIC GROWTH AND DEVELOPMENT

LEARNING OBJECTIVES

After you have studied this chapter, you should be able to

1. define economic growth, labor productivity, new growth theory, patent, and innovation;

2. identify the main determinants of economic growth;

3. indicate how economic growth affects production possibilities curves;

4. determine whether or not specific policies can contribute to economic growth;

5. evaluate the relationship between population growth and economic growth;

6. identify the fundamental stages of economic development;

7. list factors that tend to be related to a faster pace of national economic development.

CHAPTER OUTLINE

1. Economic growth is defined as the rate of increase in per-capita real GDP.
   a. Graphically, economic growth can be viewed as a rightward shift in a country's production possibilities curve.
   b. Numerically, economic growth is the sum of the rate of growth of capital, plus the rate of growth of labor, plus the rate of growth of the product of those growth rates.

2. Saving, or nonconsumption of income, is an important determinant of economic growth.
   a. When people save, they free up resources from the production of consumer goods.
   b. If such freed resources are then allocated to the production of capital or investment goods, economic growth will occur.
   c. Thus, if people consume less today, they can consume more in the future.

3. New growth theorists maintain that if technological advances are rewarded, they will be forthcoming.
   a. Patents provide property rights to intellectual achievements and inventions; thus patents encourage economic growth.

b. As long as it is profitable for people to find new ideas, they will be forthcoming, and economic growth will continue.
c. If people have incentives to increase their human capital through education and training, economic growth will be enhanced.

4. Although many people disagree, it can be shown empirically that population growth is associated with economic growth; hence immigration may be viewed as a contributing factor to economic growth.

5. When property rights are well defined, capital accumulation will ensue, and economic growth will be enhanced.

6. While it is widely believed that population growth retards a nation's economic development, there are no data to support this contention; furthermore, average family size tends to decline as a nation develops.

7. Most economically advanced nations have moved through three stages: agriculture, manufacturing, and services.

8. Several factors tend to be related to the pace of a nation's economic development: an educated population; a system of property rights; a willingness to permit new businesses to create new jobs while eliminating some old ones; and openness to international trade. A large base of natural resources is not necessarily required; natural resources must be transformed into something usable for either investment or consumption.

KEY TERMS
Economic growth
Labor productivity
New growth theory
Patent
Innovation
Development economics

COMPLETION QUESTIONS
Fill in the blank, or circle the correct term.

1. Economic growth is the _____ in per-capita real GDP.

2. Graphically, economic growth is shown as a(n) _____ shift in a nation's production possibilities curve.

3. Numerically, economic growth is equal to the _____ of the growth rate of capital and the growth rate of labor, plus _____.

4. Other things constant, if people wish to consume more in the future they must _____ more now.

5. New growth theorists predict that technology, research, and innovation (need not, must be) rewarded.

6. Inventors usually require _____ to be useful.

7. When people invest in education and training, economists refer to this as increases in human _____.

8. Economic growth is enhanced by _____, _____, _____, _____, and _____.

9. The typical stages of economic development are _____, _____, and _____.

10. A natural resource is something scarce occurring in nature that people can _____.

TRUE-FALSE QUESTIONS
Circle the **T** if the statement is true, the **F** if it is false. Explain to yourself why a statement is false.

T F 1. Economic growth generally benefits low-income people.

T F 2. Large changes in a nation's growth rate are required before significant changes in living standards can occur.

T F 3. If people save more, a nation's growth rate probably will rise.

T F 4. If an underdeveloped nation better protects property rights, its economic growth rate will rise, other things constant.

T F 5. Inventions, research, and technology occur automatically, according to the new growth theorists.

T F 6. Economic growth must eventually approach zero, because resources are finite.

T F 7. Inventions contribute to economic growth.

T F 8. When a nation invests in education, human capital increases and economic growth is enhanced.

T F 9. Underdeveloped countries can increase their economic growth by opening their economies to foreign investment.

T F 10. Empirically, immigration can be shown to reduce a nation's living standards.

T F 11. The populations of industrially advanced countries tend to grow less rapidly than those of developing countries.

T F 12. For a country to develop economically, it must have a large resource base.

MULTIPLE CHOICE QUESTIONS
Circle the letter that corresponds to the best answer.

1. Economic growth initially leads to
   a. higher living standards.
   b. increased leisure.
   c. increased pollution.
   d. All of the above

2. When economic growth occurs, the production possibilities curve
   a. shifts leftward.
   b. shifts rightward.
   c. is unaffected.
   d. rotates about the vertical axis.

3. Economic growth
   a. shifts the production possibilities curve leftward.
   b. assures full employment.
   c. may be uneven over time.
   d. creates unemployment.

4. Which of the following contributes to economic growth?
   a. technological progress
   b. well-defined property rights
   c. human capital investment
   d. All of the above

5. Before capital accumulation can take place,
   a. saving must decline.
   b. household saving must be converted into business investments.
   c. a large resource base is necessary.
   d. households must forego future consumption for more present consumption.

6. Which of the following does **NOT** contribute to economic growth?
   a. protected property rights
   b. investments in education
   c. nationalization of foreign investments
   d. increasing the ratio of investment to real GDP

7. Which of the following is **NOT** true?
   a. Saving, if converted into investment, will contribute to economic growth.
   b. Economic growth can help people of all income levels.
   c. Economic growth assures full employment.
   d. Immigration increases a nation's growth rate, other things constant.

8. Which of the following is most **UNLIKE** the others?
   a. Patents
   b. Nationalization of businesses
   c. Capital accumulation
   d. Well-defined property rights

9. Which of the following will likely increase productivity?
   a. increase in population
   b. investment in research and development
   c. higher corporate taxes
   d. elimination of patent protection

10. Which one of the following tends to be the final stage of a nation's economic development?
    a. heavy industry
    b. manufacturing
    c. agriculture
    d. services

MATCHING

Choose the item in Column (2) that best matches an item in Column (1).

(1)
a. contracts enforced by government
b. economic growth
c. inventions
d. human capital
e. saving

(2)
f. innovations
g. rate of change of per-capita real GDP
h. education
i. nonconsumption of income
j. well-defined property rights

PROBLEMS

1. Consider the hypothetical economy depicted in the following table, then answer the following questions. (Hint: each of the following questions is based on a change from one year to the next. For example, the situation in question "a" exists when going from year 1 to year 2.) Note that quantity of labor represents number of workers per year, and productivity of labor represents the real value of the output of each worker over the year.

Year	Quantity of Labor (millions)	Productivity of Labor (thousands of $)	Real GDP (billions of $)	Population (millions)	Per-Capita Real GDP (thousands of $)
1	1	$10.0	$10.0	2	$ 5.0
2	1	10.3	10.3	2	5.15
3	2	10.3	20.6	4	5.15
4	2.5	12.875	32.1875	4	8.046875
5	3.125	16.09375	50.292969	5	10.058594

a. If the quantity of labor remains constant and the productivity of labor rises by 3 percent, what happens to real GDP? Given those changes and a constant population, what happens to per-capita real GDP?

b. If the quantity of labor doubles and the productivity of labor remains constant, what happens to real GDP? Given those changes and a doubling of population, what happens to per-capita real GDP?

c. If the quantity of labor rises by 25 percent and labor productivity rises by 25 percent, what happens to real GDP? Given those changes and a constant population, what happens to per-capita real GDP?

d. If the quantity of labor rises by 25 percent, labor productivity rises by 25 percent, and population rises by 25 percent, what happens to per-capita real GDP?

ANSWERS TO CHAPTER 9

COMPLETION QUESTIONS

1. rate of increase
2. rightward
3. sum; their productivity rate of growth
4. save
5. must be
6. innovations
7. capital
8. well-defined property rights, saving, investments in human capital, open economies, population growth
9. agriculture; manufacturing; services
10. use for their own purposes

TRUE-FALSE QUESTIONS

1. T
2. F A small change in the growth rate leads to enormous changes in living standards over time.
3. T
4. T
5. F New growth theorists maintain that incentives are required to bring forth such things.
6. F As long as incentives to develop new ideas exist, economic growth will occur.
7. T
8. T
9. T
10. F Immigration and growth rates are positively correlated.
11. T
12. F Natural resources by themselves are not the key to economic development; resources must be transformed into something usable for investment or consumption.

MULTIPLE CHOICE QUESTIONS

1.d; 2.b; 3.c; 4.d; 5.b; 6.c; 7.c; 8.b; 9.b; 10. d.

MATCHING

a and j; b and g; c and f; d and h; e and i

PROBLEMS

1. a. rises by 3 percent; rises by 3 percent
   b. doubles; remains constant
   c. rises by 56.25 percent (note: this equals the sum of their changes plus their product); rises by 56.25 percent
   d. rises by 25 percent

GLOSSARY TO CHAPTER 9

Development economics The study of factors that contribute to the economic development of a country.

Economic growth Increases in per-capita real GDP measured by its rate of change per year.

Innovation Transforming an invention into something that is useful to humans.

Labor productivity Total real domestic output (real GDP) divided by the number of workers (output per worker).

New growth theory A theory of economic growth that examines the factors that determine why technology, research, innovation, and the like are undertaken, and how they interact.

Patent A government grant that gives an inventor the exclusive right or privilege to make, use, or sell an invention for a limited period of time (currently, 20 years).

CHAPTER 10
REAL GDP AND THE PRICE LEVEL IN THE LONG RUN

LEARNING OBJECTIVES
After you have studied this chapter, you should be able to

1. define aggregate demand, aggregate supply, long-run aggregate supply curve, endowments, aggregate demand curve, interest rate effect, open economy effect, and real balance effect;

2. distinguish between aggregate demand and an aggregate demand curve and between long-run aggregate supply and the long-run aggregate supply curve;

3. predict whether an aggregate demand curve will shift to the right or to the left when specific changes in non-price-level determinants occur;

4. determine the long-run equilibrium price level and the long-run equilibrium real national income level when given an aggregate demand curve and a long-run aggregate supply curve;

5. list three reasons why the aggregate demand curve is negatively sloped;

6. recognize changes that will shift the aggregate demand curve;

7. use the aggregate demand-aggregate supply approach to recognize how it is possible to have economic growth without inflation.

CHAPTER OUTLINE
1. An aggregate supply curve shows the relationship between planned rates of total production for the entire economy and various price levels.
   a. The long-run aggregate supply (LRAS) curve relates the nation's level of real national output of goods and services to the price level, when full information and full adjustments have occurred.
   b. The LRAS curve has the following properties:
      i. The LRAS curve is vertical at that level of real national output determined by tastes, technology, and the endowments of resources that exist in the nation.
      ii. The LRAS curve is vertical in the long run because a higher price level for output will be accompanied by higher costs for producers; hence, after all these adjustments are made, producers have no incentive to increase output merely because the price level is higher.

103

iii. The LRAS curve shifts rightward over time, as technological improvements occur and as the nation's endowments increase.

2. The aggregate demand curve indicates the various quantities of all goods and services demanded at various price levels.
   a. The aggregate demand curve is downward sloping for at least three reasons.
      i. When the price level rises (falls), those people who own cash balances will experience a reduction (an increase) in the purchasing power of their wealth; they consequently will plan to spend less (more) on goods and services. This is known as the real-balance effect.
      ii. When the price level rises (falls), people want to hold more (less) money in order to make the same transactions; given the supply of money, this increase (decrease) in the relative demand for money will cause interest rates to rise (fall); planned purchases on consumer durables and capital goods will therefore fall (rise). This is known as the interest rate effect.
      iii. When the price level rises (falls), domestic residents will export less (more) and import more (less); these two effects (which result from international relative price changes) cause a decrease (an increase) in planned purchases of domestically produced goods and services. This is known as the open economy effect.
   b. When the non-price-level determinants of aggregate demand change, the aggregate demand curve shifts.
      i. If the money supply rises (falls), then the AD curve will shift to the right (left); if taxes fall (rise), the AD curve will shift to the right (left).
      ii. If expectations about the future economic outlook become more (less) favorable, the AD curve will shift to the right (left).
      iii. If a nation's exchange rate decreases (increases), the AD curve will shift to the right (left).

3. The long-run equilibrium price level and the long-run equilibrium output (real national income) level are determined at the intersection of the AD curve with the LRAS curve.

4. Inflation does not necessarily accompany economic growth.
   a. Economic growth—increases in a nation's endowments of factors such as labor and capital or improvements in technology—shifts the LRAS curve rightward.
   b. In the absence of any change in aggregate demand, the price level actually would fall in a growing economy, as it did in the United States in the latter part of the nineteenth century; that is, there would be secular deflation.

5. Long-run inflation cannot result from economic growth.
   a. In the long run, inflation can result from supply-side factors only if the LRAS curve shifts leftward, which does not occur in a growing economy.
   b. Maintaining a constant long-run equilibrium price level in a growing economy requires the aggregate demand curve to shift outward at the same pace as the outward shift of the LRAS curve; hence, in the long run inflation results when the aggregate demand curve shifts rightward at a faster pace than rightward shifts in the LRAS curve.

KEY TERMS
Aggregate supply
Long-run aggregate supply
Endowments
Aggregate demand
Secular deflation

CHAPTER 10: REAL GDP AND THE PRICE LEVEL IN THE LONG RUN

KEY CONCEPTS
Long-run aggregate supply curve
Aggregate demand curve
Real-balance effect
Interest rate effect
Open economy effect

COMPLETION QUESTIONS
Fill in the blank, or circle the correct term.

1. The sum of all planned expenditures in an economy is called (aggregate demand, aggregate demand curve); the sum of planned production in the economy is called (aggregate supply, aggregate supply curve).

2. The long-run aggregate supply curve is (horizontal, vertical), because in the long run there is (full, incomplete) information, and full adjustment to changes in the price level can occur.

3. The aggregate demand curve relates planned purchase rates of all goods and services to various _____ ; the aggregate supply curve relates planned rates of total production for the entire economy to various _____ .

4. The aggregate demand curve is _____ sloped due to three effects: _____ , _____ , and _____ .

5. When the price level rises (other things being constant), the real wealth of people who hold cash balances (falls, rises); therefore planned purchases of goods and services will (fall, rise).

6. When a nation's price level falls (other things being constant) its exports (rise, fall) and its imports (rise, fall); therefore the planned purchases of its output will (rise, fall).

7. When the price level rises, people will want to hold (more, less) money in order to carry out the same transactions; this increase in the demand for money causes interest rates to (rise, fall); such a change in the interest rate causes households to plan to purchase (more, fewer) consumer durables, and businesses to plan to purchase (more, fewer) capital goods.

8. Nonprice determinants of aggregate demand include _____ , _____ , _____ , _____ , and _____ .

9. If government spending rises and taxes fall, the AD curve will shift to the (left, right); if the economic forecast is rosy, the AD curve will shift to the (left, right); if the money supply falls, the AD curve shifts to the _____ .

10. If the position of the (aggregate demand, long-run aggregate supply) curve remains unchanged, then economic growth causes the long-run equilibrium price level to (decline, increase).

TRUE-FALSE QUESTIONS

Circle the **T** if the statement is true, the **F** if it is false. Explain to yourself why a statement is false.

T F   1. The LRAS curve is vertical because firms can make adjustments and information is complete.

T F   2. The LRAS curve is vertical and doesn't shift in a growing economy.

T F   3. Aggregate demand relates planned purchases to price levels.

T F   4. The aggregate supply curve relates planned rates of total production to various price levels.

T F   5. As the price level falls, other things being constant, the purchasing power of cash balances rises.

T F   6. As the price level of a nation rises, other things being constant, the value of its imports and exports falls.

T F   7. As the price level falls, other things being constant, the demand for money falls and the interest rate rises.

T F   8. If the price level falls, the AD curve shifts to the right.

T F   9. A key factor causing the long-run equilibrium price level to rise in a growing economy is the accompanying decline in long-run aggregate supply.

T F   10. If the aggregate demand curve shifts rightward at a slower pace than rightward shifts in the LRAS curve in a growing economy, then secular deflation occurs.

MULTIPLE CHOICE QUESTIONS
Circle the letter that corresponds to the best answer.

1. Which of the following certainly will **NOT** shift the LRAS curve?
    a. a change in the price level
    b. a new oil discovery
    c. freer trade among nations
    d. economic growth

2. The LRAS curve
    a. is a short-run phenomenon.
    b. shows national output rising with the price level.
    c. does not shift over time, due to economic growth.
    d. reflects the price level/national output situation with full information and complete adjustment.

3. *Analogy*: price is to demand as the price level is to
    a. aggregate demand.
    b. aggregate supply.
    c. aggregate demand curve.
    d. aggregate supply curve.

4. Aggregate demand includes
    a. planned production rates by businesses.
    b. planned saving.
    c. planned purchases by households and businesses.
    d. various price levels.

5. The aggregate demand curve
   a. relates planned purchases to various price levels.
   b. is negatively sloped.
   c. includes planned expenditures on consumption, investment, and governmentally provided goods.
   d. All of the above

6. The aggregate demand curve is negatively sloped because, other things being constant,
   a. as the price level rises, the demand for money falls.
   b. as the price level falls, the purchasing power of cash balances rises.
   c. as the price level falls, the AD curve shifts to the right.
   d. as the price level rises, exports rise.

7. When the price level falls, other things being constant,
   a. the demand for money falls.
   b. the interest rate falls.
   c. household and business planned expenditures rise.
   d. All of the above

8. Which of the following does **NOT** occur when the price level rises, other things being constant?
   a. Exports rise and imports fall.
   b. The demand for money rises.
   c. The purchasing power of cash balances falls.
   d. Aggregate demand falls.

9. Which of the following does **NOT** cause the AD curve to shift?
   a. an increase in taxes
   b. a decrease in the real interest rate
   c. a decrease in the price level
   d. a change in the money supply

10. Which of the following does **NOT** cause secular deflation?
    a. economic growth
    b. a decrease in long-run aggregate supply
    c. a decrease in aggregate demand at a faster pace than a decrease in long-run aggregate supply
    d. failure of aggregate demand to increase in the face of an increase in long-run aggregate supply

MATCHING
Choose the item in Column (2) that best matches an item in Column (1).

(1)

a. aggregate demand
b. aggregate supply
c. aggregate demand shift
d. real-balance effect
e. open economy effects
f. interest rate effects

(2)

g. change in government spending or taxing
h. changes in imports and exports due to price changes
i. planned production
j. changes in the demand for money due to changes in the price level
k. planned expenditures
l. change in the value of cash balances

WORKING WITH GRAPHS

1. Consider the graph below, and then answer the following questions.

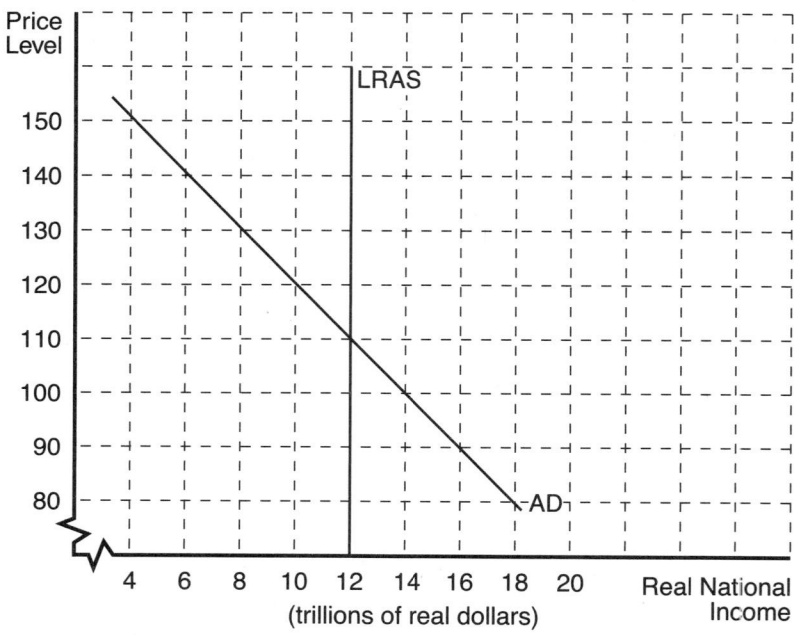

   a. What is the current long-run equilibrium level of real national income? What is the current long-run equilibrium price level?
   b. If the economy grows sufficiently that $2 trillion in additional real national income is forthcoming in the long run, and if aggregate demand remains unchanged, what will be the new long-run equilibrium price level?

2. Consider the diagram below, and suppose that the long-run equilibrium level of real national income rises by $2 trillion, but the equilibrium price level remains unchanged. Assuming parallel shift(s) of any schedule, draw new schedules showing how this could take place.

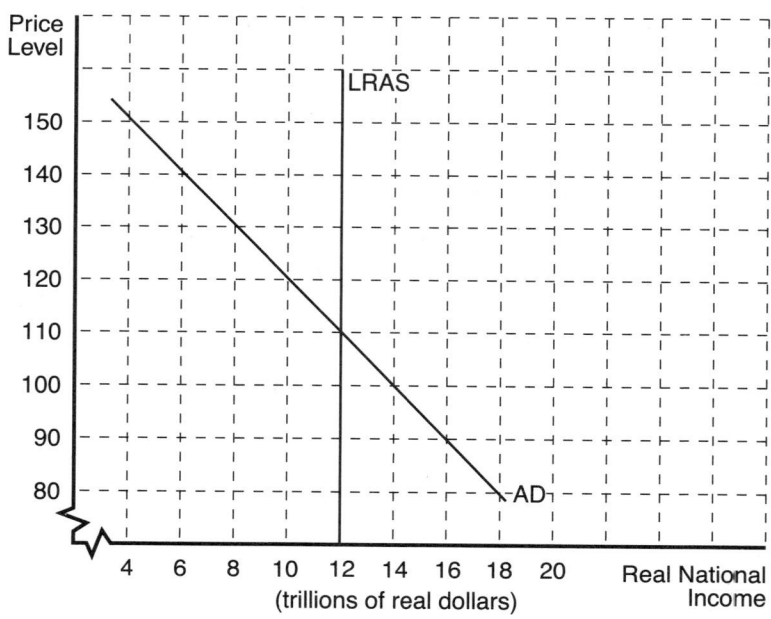

ANSWERS TO CHAPTER 10

COMPLETION QUESTIONS
1. aggregate demand; aggregate supply
2. vertical; full
3. price levels; price levels
4. negatively; real balance, interest rate, substitution of foreign goods
5. falls; fall
6. rise, fall; rise
7. more; rise; fewer, fewer
8. government spending and taxing policies, exchange rates, expectations, money supply, real interest rates
9. right; right; left
10. aggregate demand; decline

TRUE-FALSE QUESTIONS
1. T
2. F It shifts rightward in a growing economy.
3. F The AD *curve* relates planned purchases to price levels.
4. T
5. T
6. F The value of exports falls, imports rise.
7. F Interest rates fall.
8. F A lower price level leads to a movement down the AD curve; no shift.
9. F In a growing economy, aggregate supply increases, so the LRAS curve shifts rightward.
10. T

MULTIPLE CHOICE QUESTIONS
1.a;   2.d;   3.a;   4.c;   5.d;   6.b;   7.d;   8.a;   9.c;   10.b.

MATCHING
a and k; b and i; c and g; d and l; e and h; f and j

WORKING WITH GRAPHS

1.
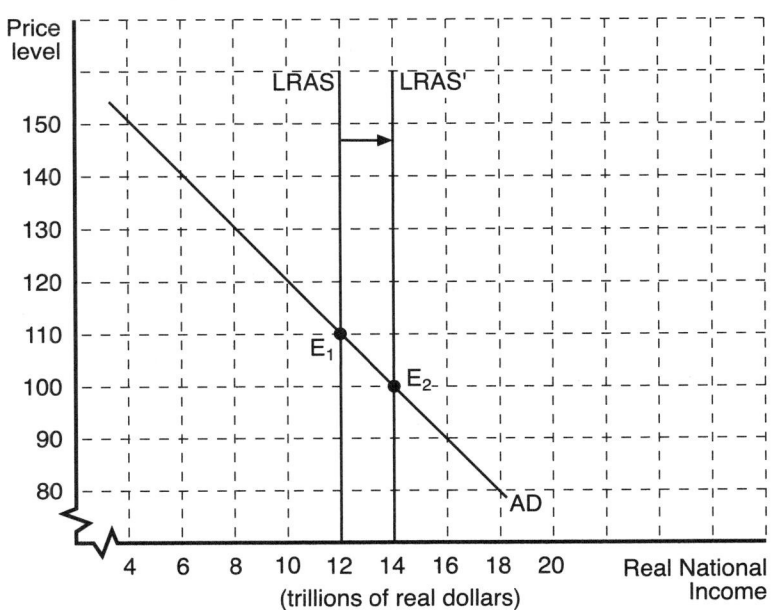

a. The initial long-run equilibrium level of real national income at point E_1 is 12 trillion real dollars, and the initial long-run equilibrium price level is 110.
b. If the economy grows sufficiently that $2 trillion in additional real national income is forthcoming in the long run, then the long-run aggregate supply curve shifts to the right by this amount, and the new long-run equilibrium price level falls to 100 at point E_2. There is secular deflation.

2. If the long-run equilibrium level of real national income rises by $2 trillion, then the long-run aggregate supply curve shifts rightward by this amount. The equilibrium price level can remain unchanged at point E_2 only if the aggregate demand curve shifts up sufficiently, as shown, to prevent secular deflation from occurring.

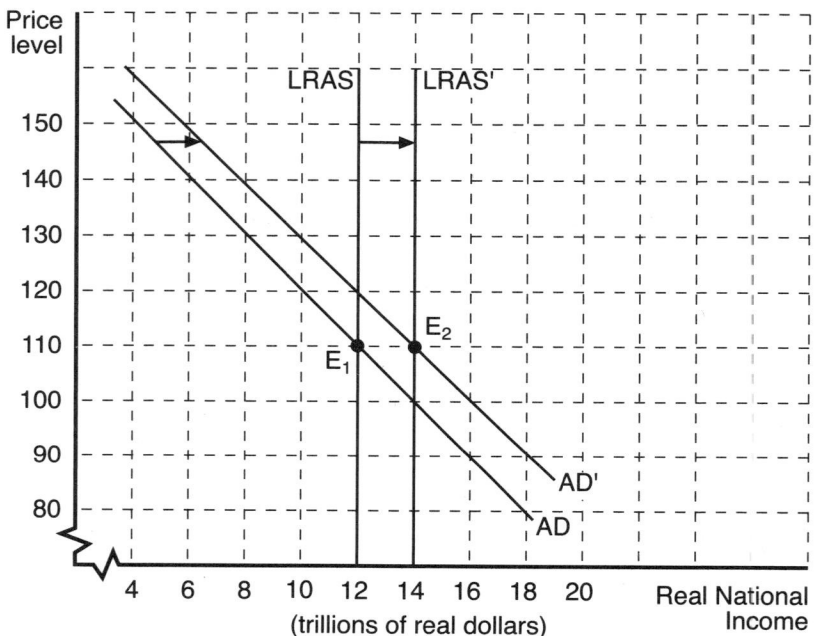

GLOSSARY TO CHAPTER 10

Aggregate demand The total of all planned expenditures for the entire economy.

Aggregate demand curve A curve showing planned purchase rates for all goods and services in the economy at various price levels, all other things held constant.

Aggregate supply The total of all planned production for the economy.

Endowments The various resources in an economy, including both physical resources and such human resources as ingenuity and management skills.

Interest rate effect One of the reasons that the aggregate demand curve slopes downward: Higher price levels increase the interest rate, which in turn causes businesses and consumers to reduce desired spending due to the higher price of borrowing.

Long-run aggregate supply curve A vertical line representing the real output of goods and services after full adjustment has occurred. It can also be viewed as representing the real output of the economy under conditions of full employment—the full-employment level of real GDP.

Open economy effect One of the reasons that the aggregate demand curve slopes downward: Higher price levels result in foreign residents desiring to buy fewer U.S.-made goods, while U.S. residents now desire more foreign-made goods, thereby reducing net exports. This is equivalent to a reduction in the amount of real goods and services purchased in the United States.

Real-balance effect The change in expenditures resulting from a change in the real value of money balances when the price level changes, all things held constant; also called the *wealth effect*.

Secular deflation A persistent decline in prices resulting from economic growth in the presence of stable aggregate demand.

CHAPTER 11
CLASSICAL AND KEYNESIAN MACRO ANALYSIS

LEARNING OBJECTIVES

After you have studied this chapter, you should be able to

1. define Say's law, money illusion, Keynesian short-run aggregate supply curve, aggregate demand shock, recessionary gap, inflationary gap, demand-pull inflation, and cost-push inflation;

2. recognize the main assumptions and conclusions of the classical model;

3. recognize the shape of the classical long-run aggregate supply curve, and predict the effects of a change in aggregate demand in this model;

4. recognize the shape of the Keynesian short-run aggregate supply curve and predict the effects of a change in aggregate demand in this model;

5. recognize reasons why the short-run aggregate supply curve is positively sloped;

6. distinguish between the short-run and the long-run aggregate supply curves;

7. predict the effects on the aggregate supply curve and the aggregate demand curve of a change in a nation's exchange rate;

8. distinguish between a recessionary gap and an inflationary gap;

9. distinguish between cost-push and demand-pull inflation.

CHAPTER OUTLINE

1. The classical model was developed in the 18th and 19th centuries; it attempted to explain the determinants of the price level and GDP.
   a. Say's law is that supply creates its own demand.
   b. The classical model assumes (1) pure competition, (2) wage and price flexibility, (3) self-interest, and (4) no money illusion.

112

CHAPTER 11: CLASSICAL AND KEYNESIAN MACRO ANALYSIS 113

    c.   In this model the interest rate changes until household saving is equated to business planned investment, thereby assuring that every dollar that leaves the economic system as saving comes back in as a dollar invested.
    d.   Because the LRAS curve is vertical in the classical model, changes in AD merely change the price level; GDP is supply-determined.

2. John Maynard Keynes developed what is now the Keynesian model during the 1930s, a depression period.
    a.   This model assumes that prices and wages are constant in the short run; hence the SRAS curve has a horizontal range.
    b.   Changes in AD, consequently, lead to changes in GDP, but not the price level; in this model GDP is demand-determined in the short run.

3. The short-run aggregate supply (SRAS) curve shows how real national output changes with the price level in the short run, before full adjustment is made and before full information is available.
    a.   If the price level rises while input costs remain the same, producers have an incentive to increase output; if prices rise and some costs remain constant, it is profitable to increase output.
    b.   Producers are able to expand output in the short run in response to price level increases because
        i.   firms can use existing workers more intensively.
        ii.   existing capital equipment can be used more intensively.
        iii.   if wage rates don't rise, it is profitable for firms to hire the unemployed or new entrants to the labor force.
    c.   At some point the SRAS curve becomes very steep; it becomes harder and harder to find more workers at existing wage rates.
    d.   The SRAS curve, therefore, is positively sloped and becomes steeper and steeper as the price level gets higher.
    e.   Changes in the non-price-level determinants of short-run aggregate supply lead to shifts in the short-run aggregate supply curve.
        i.   If input costs rise (fall), the SRAS curve shifts to the left (right).
        ii.   An increase (decrease) in labor or in its productivity will shift the SRAS curve to the right (left).
        iii.   Temporary changes in the above shift SRAS, but not LRAS.

4. Unanticipated shifts in the AD and the AS curves are called aggregate demand shocks and aggregate supply shocks, respectively.
    a.   A short war, such as the Persian Gulf War, shifts the AD curve rightward; the equilibrium price level rises and the equilibrium output level rises.
    b.   If the war is prolonged, the SRAS curve shifts upward, and the new equilibrium position is at a higher price level, but at the old, initial, national output level.

5. If full employment equilibrium does not exist, then either an inflationary gap or a recessionary gap exists.
    a.   An inflationary gap exists when the equilibrium level of real national income is greater than the full employment level.
    b.   A recessionary gap exists when the equilibrium level of real national income is less than the full employment level.

6. There are two types of inflation—demand-pull and cost-push.
    a.   Demand-pull inflation is caused by increases in aggregate demand that are not matched by increases in aggregate supply.

114 CHAPTER 11: CLASSICAL AND KEYNESIAN MACRO ANALYSIS

    b. Cost-push inflation is caused by continual leftward shifts in the short-run aggregate supply curve.

7. If a nation's exchange rate rises in value, its SRAS curve will shift rightward, as imported raw material prices fall, and its AD curve will shift leftward as its imports rise and its exports fall; the net effect is a lower price level, while the net effect on GDP is indeterminate.

KEY TERMS
Say's law   Recessionary gap   Demand-pull inflation
Short-run aggregate supply   Inflationary gap   Cost-push inflation

KEY CONCEPTS
Money illusion   Keynesian short-run aggregate supply curve
Aggregate supply shock   Aggregate demand shock

COMPLETION QUESTIONS
Fill in the blank or circle the correct term.

1. Say's Law maintains that _____ creates its own _____.

2. The major assumptions of the classical model are _____, _____, _____, and _____.

3. If people respond to changes in absolute prices when relative prices are unaltered, then they suffer from a(n) _____ illusion.

4. In the classical model desired household saving and desired business investment are equated by the _____ rate; and at the equilibrium wage rate, full employment (does, does not) exist.

5. In the classical model the LRAS curve is (horizontal, vertical); therefore changes in the AD curve lead to changes in _____, but not to changes in _____.

6. In the Keynesian range, prices and wage rates are (fixed, flexible), hence the SRAS curve is (horizontal, vertical); changes in AD lead to changes in _____, but not in _____.

7. Saving is (a leakage from, an injection into) the circular flow and is a potential problem; if saving is offset by _____, which is (a leakage from, an injection into) the circular flow, then full employment will prevail.

8. Saving represents a(n) _____ curve; investment represents a(n) _____ curve. Saving and investment are brought into equality by the _____ rate, in the classical model.

9. If much unused capacity and massive unemployment exist, the economy will be operating in the _____ range of the SRAS curve; in that range, increases in the AD curve will lead to increases in _____ but no changes in the _____; in that range, national income is said to be _____ determined.

We relax the assumption of a fixed price level in questions 10, 11, 12, and 13.

10. An increase in AD causes the price level to _____, but some of that effect is reduced because of the following effects: _____, _____, and _____.

11. A decrease in AD causes the price level to _____, but some of the effect is reduced.

12. Nation A's exchange rate has increased. As a result its SRAS curve will shift to the _____, and its AD curve will shift to the _____; the net effect is that nation A's price level will _____ and its GDP change will _____.

13. Country Z's exchange rate has weakened. As a result its AD curve will _____ and its SRAS curve will _____; the net effect on country Z is that its GDP change will _____ and its price level will _____.

14. As economic growth occurs, a nation's short-run AS curve will shift _____ and its long-run AS curve will shift _____. Consequently, if the AD curve shifts rightward during periods of economic growth, inflation (will, will not, may not) result.

15. The short-run aggregate supply curve is _____ sloped; as the price level rises in the short run, planned production by businesses (rises, falls).

16. If changes in production costs lag behind changes in the price level, producers will have an incentive to produce (more, less) as the price level rises.

17. The short-run AS curve will shift if there is a change in any of the following non-price-level determinants of aggregate supply: _____, _____, _____, and _____.

18. If wage rates fall, the short-run AS curve will shift to the _____; if technological improvements occur and if the prices of raw materials fall permanently, the long-run AS curve will shift to the _____.

19. The price level and the equilibrium national output level are determined where the SRAS curve and the AD curve _____.

20. At very high price levels, the SRAS curve becomes very (flat, steep) because it becomes (more difficult, easier) to get more labor at relatively fixed wage rates.

21. A temporary increase in an input price shifts only the (LRAS, SRAS).

22. An unanticipated shift in the AD curve is called a(n) _____.

23. If equilibrium does not exist, then either a(n) _____ or a(n) _____ exists.

24. Continual leftward shifts in the SRAS cause _____.

TRUE-FALSE QUESTIONS
Circle the **T** if the statement is true, the **F** if it is false. Explain to yourself why a statement is false.

T F   1.   The classical model preceded the Keynesian model.

T F   2.   Say's Law says that demand creates its own supply.

T F   3.   In the classical model, prices and wages are fixed.

T F 4. A money illusion exists if people respond to relative, not absolute, price or wage rate changes.

T F 5. In the classical model, household desired saving equals business desired investment because the interest rate adjusts until they are equated.

T F 6. The classical model is consistent with the horizontal range of the SRAS curve.

T F 7. In the horizontal range of the SRAS curve, real GDP is demand-determined.

T F 8. If the AD curve shifts in the classical model, then the price level will change, but the level of real national output remains constant.

T F 9. If the AD curve shifts in the Keynesian range, the price level changes, as does real GDP.

T F 10. If the interest rate falls, investment spending will rise.

T F 11. If the price level can change, a rightward shift in AD will cause real GDP to rise and the price level to fall.

T F 12. If a nation's exchange rate weakens, its SRAS curve will shift leftward, its AD curve will shift rightward, and its price level will rise.

T F 13. Economic growth causes a nation's long-run AS curve to shift to the right, but its short-run AS curve to shift to the left.

T F 14. If the price level rises and the costs of inputs don't rise immediately, producers have an incentive to increase output.

T F 15. The short-run aggregate supply curve is positively sloped.

T F 16. If productivity rises and raw material prices fall, then the SRAS curve will shift to the right.

T F 17. If the price level falls and wage rates don't, producers have an incentive to produce less.

T F 18. Producers can expand output in the short run by adding to their capital stock.

T F 19. The SRAS curve becomes very flat at higher and higher price levels.

T F 20. A demand shock that shifts the AD curve rightward will probably cause national output to rise and the price level to fall.

T F 21. If neither a recessionary nor an inflationary gap exists, then full employment equilibrium exists.

T F 22. Cost-push inflation involves continual shifts inward in the short-run aggregate supply curve.

MULTIPLE CHOICE QUESTIONS
Circle the letter that corresponds to the best answer.

1. The classical model assumes
    a. prices and wages are constant.
    b. pure competition.
    c. people suffer from money illusion.
    d. altruism is the main motivating force.

2. Say's Law
    a. was developed by J. M. Keynes.
    b. maintains that demand creates its own supply.
    c. maintains that supply creates its own demand.
    d. implies that general overproduction is likely under capitalism.

3. In the classical model, saving
    a. is an injection into the income stream.
    b. and investment are determined by national disposable income.
    c. causes unemployment.
    d. is a leakage from the circular flow.

4. In the classical model, when households save,
    a. that money becomes a part of the supply of saving curve.
    b. business investment offsets such saving.
    c. full employment will still prevail.
    d. All of the above

5. In the Keynesian model, in a depression
    a. the economy operates in the horizontal range of the SRAS curve.
    b. prices and wages are flexible.
    c. the interest rate adjusts until desired saving equals desired investment.
    d. All of the above

6. When economic growth occurs, a nation's
    a. short-run AS curve shifts leftward.
    b. short-run and long-run AS curves shift rightward.
    c. long-run AS curve becomes horizontal.
    d. price level must rise.

7. In the horizontal range of the SRAS curve,
    a. the price level is constant.
    b. the classical model assumptions apply.
    c. national income is supply-determined.
    d. full employment exists.

8. If a nation's exchange rate rises, its
    a. SRAS curve shifts leftward, due to input cost reductions.
    b. AD curve shifts rightward, due to falling imports.
    c. price level will fall.
    d. GDP will rise, unambiguously.

9. If a nation's exchange rate falls, its
   a. SRAS curve shifts leftward, due to input cost increases.
   b. AD curve shifts rightward, due to increased exports and decreased imports.
   c. price level rises.
   d. All of the above

10. Which of the following will **NOT** shift the classical investment curve?
    a. a change in the interest rate
    b. a change in profit expectations
    c. a change in business taxes
    d. a change in technology and innovation

11. In the classical model, if AD shifts to the right, then
    a. real national income rises.
    b. the price level rises.
    c. real national income falls.
    d. the price level is unaffected.

12. In the Keynesian range of SRAS, if AD shifts to the left, then
    a. the price level falls.
    b. real national income is unchanged.
    c. real national income falls.
    d. the price level rises.

13. Concerning the SRAS curve, an increase in the price level
    a. has no effect on planned production.
    b. leads to an increase in output if input prices rise proportionally.
    c. causes a decrease in planned production in the long run.
    d. causes an increase in planned production.

14. If the price level rises faster than costs of production rise, then
    a. profits per unit fall.
    b. producers have an incentive to increase output.
    c. the aggregate supply curve shifts to the right.
    d. the aggregate supply curve shifts to the left.

15. Which of the following causes the SRAS curve to shift to the left?
    a. a fall in wage rates
    b. rises in productivity
    c. technological improvements
    d. increase in raw material costs

16. If unused capacity and significant unemployment exist, then the SRAS curve is
    a. vertical over a broad range.
    b. downward sloping.
    c. horizontal or slightly upward sloping.
    d. the same as the LRAS curve.

17. At the intersection of the SRAS curve, LRAS curve, and the AD curve,
    a. the equilibrium price level is determined.
    b. the equilibrium national output level is determined.
    c. economywide equilibrium exists.
    d. All of the above

18. An unanticipated rightward shift in the AD curve
    a. is a supply shock.
    b. is a demand shock.
    c. will cause the output level to fall.
    d. will cause the price level to fall.

19. Massive technological improvements in the computer industry probably will cause the
    a. SRAS curve to shift upward.
    b. LRAS curve to shift rightward.
    c. AD curve to shift leftward.
    d. AD curve to shift rightward because the price level will rise.

20. Both unemployment and the price level rise if the
    a. AD curve shifts to the right.
    b. AD curve shifts to the left.
    c. SRAS curve shifts to the left.
    d. SRAS curve shifts to the right.

21. A temporary rise in production costs
    a. shifts the LRAS curve rightward.
    b. shifts the SRAS curve upward.
    c. shifts the AD curve leftward.
    d. shifts the AD curve rightward.

22. Which of the following is most **UNLIKE** the others?
    a. rosy economic outlook
    b. rise in a nation's exchange rate
    c. tax decreases
    d. increase in money supply

MATCHING
Choose the item in Column (2) that best matches an item in Column (1).

(1)	(2)
a. Say's law	i. confusion between absolute and relative prices
b. fixed price level	j. leftward shift in AD
c. money illusion	k. Keynesian range
d. stronger exchange rate	l. classical model
e. weaker exchange rate	m. leftward shift in SRAS
f. economic growth	n. expenditure on capital goods
g. investment	o. rightward shifts in SRAS and LRAS
h. aggregate demand shock	p. unanticipated change in investment expenditures

WORKING WITH GRAPHS

1. Consider the short-run graphs below, then answer the questions that follow.

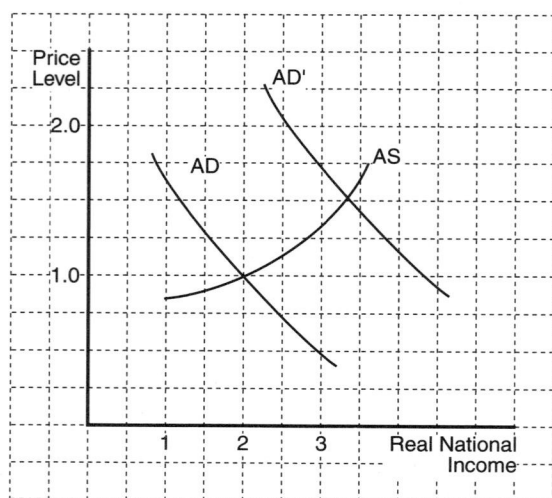

   a. What is the equilibrium price level, given the AS and AD curves? What is the equilibrium level of real national income?
   b. What could cause the AD curve to shift to AD'? Answer the remaining questions assuming that AD has shifted to AD'.
   c. Given AS and the previous AD curve, what now happens to the price level? Why?
   d. Given AS and the previous AD curve, what now happens to the equilibrium level of real national income? Why?
   e. If full employment exists at a real national income level of $2, what exists at the new equilibrium level?

2. Using the coordinate system below, answer the questions that follow.

   a. Assuming a flexible price level and an open economy, draw an equilibrium situation using the AS/AD model.
   b. Assume that this nation's exchange rate has weakened. Draw in the new AD and the new SRAS curves.
   c. What happens to GDP and the price level?

PROBLEMS

1. Assuming the economy is operating at equilibrium, predict what happens to the equilibrium price level and the equilibrium real national income level as a result of
   a. productivity increases;
   b. an increase in the labor participation rates;
   c. a significantly lower marginal tax rate;
   d. economic growth;
   e. a temporary rise in raw material prices;
   f. an increase in population;
   g. a decrease in raw material prices;
   h. a decrease in government spending;
   i. an improvement in expectations about the future economic outlook;
   j. an increase in the price of oil;
   k. technological improvements;
   l. a temporary increase in investment spending;
   m. a permanent increase in investment spending.

ANSWERS TO CHAPTER 11

COMPLETION QUESTIONS

1. supply; demand
2. pure competition; price-wage flexibility; self-interest; no money illusion
3. money
4. interest; does
5. vertical; the price level; real GDP
6. fixed; horizontal; real GDP; the price level
7. a leakage from; investment; an injection into
8. supply; demand; interest
9. horizontal; real GDP; price level; demand
10. rise; wealth; interest rate; foreign goods substitution effects
11. fall
12. right; left; fall; be indeterminate
13. shift rightward; shift leftward; be indeterminate; rise
14. rightward; rightward; may not
15. positively; rises
16. more
17. wage rates, productivity, technology, raw material prices
18. right; right
19. intersect
20. steep; more difficult
21. SRAS
22. aggregate demand shock
23. recessionary gap; inflationary gap
24. cost-push inflation

TRUE-FALSE QUESTIONS

1. T
2. F Say's law is that supply creates its own demand.
3. F They are flexible in the classical model.
4. F Money illusion results when people respond to absolute, but not relative, price changes.
5. T
6. F The SRAS curve is vertical in the classical model.
7. T
8. T
9. F Price level is constant; real GDP changes.
10. T
11. F Real GDP rises, but the price level rises too.
12. T
13. F Both curves shift rightward.
14. T
15. T
16. T
17. T
18. F By assumption, this option is not available in the short run.
19. F It becomes very steep as labor becomes more difficult to obtain at constant wage rates.
20. F The price level will rise too.
21. T
22. T

MULTIPLE CHOICE QUESTIONS

1.b; 2.c; 3.d; 4.d; 5.a; 6.b; 7.a; 8.c; 9.d; 10.a; 11.b; 12.c; 13.d; 14.b; 15.d; 16.c; 17.d; 18.b; 19.b; 20.c; 21.b; 22.b.

CHAPTER 11: CLASSICAL AND KEYNESIAN MACRO ANALYSIS

MATCHING

a and l; b and k; c and i; d and j; e and m; f and o; g and n; h and p

WORKING WITH GRAPHS

1. a. 1.0, 2
   b. Increase in government expenditures, decrease in taxes, increase in money supply, increase in population, rosier expectations
   c. Rises, because a general shortage of goods and services now exists at the previous price level.
   d. Rises, because producers have an incentive to produce more as the price level rises.
   e. Inflationary gap

2. a. N.A.
   b. The AD curve should shift rightward; the AS curve should shift leftward.
   c. Effect on GDP is indeterminate; price level should rise.

PROBLEMS

1. a. The SRAS and LRAS curves shift to the right; therefore national output rises, and the price level falls.
   b. The SRAS curve shifts to the right; therefore national output rises, and the price level falls.
   c. The SRAS curve shifts to the right; therefore national output rises, and the price level falls.
   d. The LRAS curve (and SRAS curve) shifts to the right as potential output increases; national output rises, and the price level falls.
   e. Temporary rise in price level; temporary reduction in national output.
   f. The AD curve shifts to the right; therefore national output and the price level rise.
   g. The SRAS curve shifts to the right; therefore the price level falls and national output rises.
   h. The AD curve shifts to the left; national output falls and the price level falls.
   i. The AD curve shifts to the right; therefore national output rises, and the price level rises.
   j. The SRAS curve shifts to the left; therefore national output falls, and the price level rises.
   k. The SRAS and LRAS curves shift to the right; therefore national output rises, and the price level falls.
   l. The AD curve shifts to the right; therefore national output and the price level rise in the short run.
   m. LRAS shifts as well as AD; national output rises, but the effect on the price level is uncertain.

GLOSSARY TO CHAPTER 11

Aggregate demand shock Any event that causes the aggregate demand curve to shift inward or outward.

Aggregate supply shock Any event that causes the aggregate supply curve to shift inward or outward.

Cost-push inflation Inflation caused by a continually decreasing short-run aggregate supply curve.

Demand-pull inflation Inflation caused by increases in aggregate demand not matched by increases in aggregate supply.

Inflationary gap The gap that exists whenever the equilibrium level of real national income per year is greater than the full-employment level as shown by the position of the long-run aggregate supply curve.

Keynesian short-run aggregate supply curve The horizontal portion of the aggregate supply curve in which there is unemployment and unused capacity in the economy.

Money illusion Reacting to changes in money prices rather than relative prices. If a worker whose wages double when the price level also doubles thinks he or she is better off, that worker is suffering from money illusion.

Recessionary gap The gap that exists whenever the equilibrium level of real national income per year is less than the full-employment level as shown by the position of the long-run aggregate supply curve.

Say's law A dictum of economist J.B. Say that supply creates its own demand; producing goods and services generates the means and the willingness to purchase other goods and services.

Short-run aggregate supply curve The relationship between total planned production and the price level in the short run, all other things held constant. If prices adjust gradually in the short run, the curve is positively sloped.

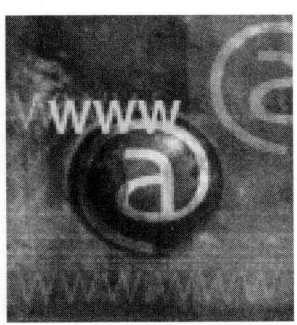

CHAPTER 12
CONSUMPTION, INCOME, AND THE MULTIPLIER

LEARNING OBJECTIVES

After you have studied this chapter, you should be able to

1. define saving, savings, consumption, dissaving, autonomous consumption, average propensity to consume, average propensity to save, marginal propensity to consume, marginal propensity to save, 45-degree reference line, wealth, lump-sum tax, and multiplier;

2. distinguish between flow variables and stock variables;

3. relate consumption and saving to income;

4. distinguish between the marginal propensity to consume (save) and the average propensity to consume (save);

5. calculate marginal and average propensities to consume (save), given the relevant information;

6. distinguish between the causes of a movement along and a shift in the consumption (saving) curve;

7. list the main determinants of planned business investment spending and recognize how each affects such spending;

8. predict what will happen to national income and employment if total planned expenditures do not equal real national income;

9. calculate the autonomous consumption or autonomous investment multiplier, given the relevant information;

10. calculate the change in the equilibrium level of national income due to a change in autonomous expenditures, given the marginal propensity to consume.

CHAPTER 12: CONSUMPTION, INCOME, AND THE MULTIPLIER

CHAPTER OUTLINE

1. The sum of consumption expenditures and saving equals disposable income, by definition.
   a. Saving, consumption, and income are flows which, therefore, are measured per unit of time; savings and wealth are stocks whose values are measured at a given moment in time.
   b. Investment, which is a flow, includes expenditures by firms for capital goods.

2. John Maynard Keynes maintained that planned real saving and planned real consumption were determined by real disposable income.
   a. For the household, planned real consumption is typically plotted on the vertical axis and real disposable income on the horizontal axis; the planned saving curve can be derived by subtracting the planned consumption curve from the 45 degree line.
   b. Autonomous consumption is that consumption that is independent of income, and it is the value of the vertical intercept on the consumption function; if consumption is positive at zero national income, then dissaving must exist over the lower range of the consumption function.
   c. The average propensity to consume (APC) equals total real consumption divided by total real disposable income; the average propensity to save (APS) equals total real saving divided by total real disposable income.
   d. The marginal propensity to consume (MPC) equals the change in real consumption divided by the change in real disposable income; the marginal propensity to save (MPS) equals the change in real saving divided by the change in real disposable income.
   e. The APC + APS = 1; and the MPC + MPS = 1.
   f. When real disposable income changes, a movement along the consumption curve results; when there is a change in the nonincome determinants of consumption, the consumption curve shifts.

3. The nonincome determinants of consumption include wealth, future income or inflationary expectations, and population.
   a. If wealth increases, the consumption function shifts upward; if wealth decreases, it shifts downward.
   b. If households expect better times ahead, the consumption function shifts upward; expected worse times will shift it downward.
   c. An expectation that the inflation rate will fall may shift the consumption function downward.
   d. An increase in population shifts the consumption function upward; a decrease in population shifts it downward.

4. Investment includes business expenditures on plants and equipment and on inventories; such expenditures are more variable than household consumption expenditures because expectations play a more important role in investment expenditures.
   a. An inverse relationship exists between investment expenditures and the interest rate.
   b. Noninterest rate determinants of investment spending include expectations, innovation and technology, and business taxes.
      i. If the future looks rosier, then more investment expenditures will be made at any interest rate; the investment curve will shift to the right.
      ii. Improvements in technology and innovation shift the planned investment curve to the right.
      iii. If business taxes rise, the planned investment curve shifts to the left; a reduction in such taxes shifts it to the right.

5. The simplified Keynesian model, which considers only household consumption expenditures and business investment expenditures, is useful because it provides insights and serves as a foundation for more complicated models.

6. Real consumption depends on real disposable income and it is also related to real national income; the latter relationship is more convenient for our analysis.
   a. A portion of the consumption function is autonomous, or independent of real national income.
   b. Net investment is dependent on the interest rate; for simplicity we assume that it is autonomous, or independent of national income.
   c. The 45-degree reference line indicates where planned expenditures equal real national income per year.

7. The equilibrium level of real national income occurs at that income level where the planned expenditures curve intersects the 45-degree reference line.
   a. Once the equilibrium real national income level is determined, the employment level is also determined because a functional relationship exists between those two variables in the short run.
   b. At all other income levels, disequilibrium exists.
      i. If total planned expenditures exceed real national income, then business inventories will fall involuntarily and businesses will find it profitable to increase output and employment.
      ii. If total planned expenditures are less than real national income, then business inventories will rise involuntarily and businesses will find it profitable to decrease output and employment.

8. When autonomous expenditures change, the planned expenditures curve shifts and there will be a multiplier effect.
   a. If autonomous consumption, autonomous investment, autonomous government expenditures, or net exports change, the planned expenditures curve shifts by an identical amount.
   b. Equilibrium real national income per year will change by a multiple of the change in autonomous expenditure, in the same direction.

9. A multiplier effect exists because one person's expenditure is another person's income, and changes in autonomous spending lead to successive rounds of spending and income creation.
   a. The steeper the slope of the planned expenditures curve (the MPC), the greater is the multiplier.
   b. The simple multiplier equals the reciprocal of the MPS.
   c. Because of the multiplier effect, fluctuations in economic activity will be magnified.

KEY TERMS
45-degree reference line
Saving
Wealth
Dissaving
Consumption
Average propensity to consume
Marginal propensity to consume
Average propensity to save
Marginal propensity to save

KEY CONCEPTS
Autonomous consumption
Multiplier
Lump-sum tax
Consumption function
Consumption goods
Capital goods

COMPLETION QUESTIONS
Fill in the blank, or circle the correct term.

1. Saving is a (stock, flow) concept, while savings is a(n) _____ concept.

2. The sum of planned consumption and planned saving equals _____, by definition; when real disposable income rises, planned real consumption _____ and planned real saving _____.

3. The average propensity to save (APS) equals saving divided by _____; the APC plus the APS equals _____.

4. The marginal propensity to consume (MPC) equals the change in consumption (divided by, plus, minus) the change in real disposable income; one minus the MPC equals the _____.

5. The amount of consumption that is (dependent on, independent of) income is called autonomous consumption; when autonomous consumption exists, the vertical intercept of the consumption function is (negative, zero, positive), and the APC (falls, remains constant, rises) as real disposable income rises.

6. Dissaving exists when consumption expenditures (equal, are less than, exceed) income.

7. The consumption function will shift if autonomous consumption changes due to changes in such nonincome determinants of consumption as _____, _____, and _____.

8. Investment varies (directly, inversely) with changes in the interest rate; the planned investment curve shifts if there are changes in _____, _____, or _____.

9. Along the 45-degree reference line, planned total expenditures equal real _____; where the planned expenditures line intersects the 45-degree reference line (equilibrium, disequilibrium) exists; where those curves do not intersect, _____ exists.

10. In the model in which government and foreign transactions are ignored, household planned consumption expenditures plus business investment expenditures equal aggregate _____.

11. If total planned expenditures exceed real national income, business inventories will _____ involuntarily and businesses will find it profitable to (increase, decrease) output and employment; if total planned expenditures are less than real national income, business inventories will _____ involuntarily and businesses will find it profitable to _____ output and employment.

12. If autonomous government purchases of goods and services (G) are added to the aggregate expenditures curve, the aggregate expenditures curve will shift (upward, downward) and equilibrium real national income per year will (rise, fall).

13. If net exports are added to the aggregate expenditure curve and imports exceed exports, then net exports are a (positive, negative) number and equilibrium real national income will (rise, fall) by an amount (greater than, equal to, less than) net exports.

TRUE-FALSE QUESTIONS
Circle the **T** if the statement is true, the **F** if it is false. Explain to yourself why a statement is false.

T F 1. The APC plus the MPC equals 1, by definition.

T F 2. In the Keynesian model, if wealth rises, the consumption function shifts upward.

T F 3. In the Keynesian model, the APC falls and the APS rises as national income rises.

T F 4. If autonomous consumption is positive, then the vertical intercept of the consumption curve is positive.

T F 5. If real disposable income rises, the consumption function will shift upward.

T F 6. On average, disposable income is about 85 percent of national income.

T F 7. The 45-degree reference line indicates planned expenditures at each level of real national income.

T F 8. Autonomous consumption and autonomous investment vary directly with real national income.

T F 9. The equilibrium level of real national income is found at the point at which the planned expenditures curve intersects the 45-degree reference line.

T F 10. If total planned expenditures exceed real national income, then business inventories will fall and businesses will increase output.

T F 11. In the short run, employment and real national income (output) are directly related.

T F 12. Ignoring the government and foreign sectors, if planned saving is less than planned investment, then planned expenditures are less than real national income.

T F 13. If autonomous expenditures rise, then the planned expenditures curve will shift upward.

T F 14. If autonomous expenditures rise by $1 billion, national income will probably rise by more than $1 billion.

T F 15. If the price level falls, the planned expenditures curve will shift upward.

T F 16. If the MPC is 0.75, a $1 billion increase in autonomous expenditures will cause national income to rise by $4 billion, if the SRAS curve is horizontal.

MULTIPLE CHOICE QUESTIONS
Circle the letter that corresponds to the best answer.

1. Autonomous real consumption
   a. varies directly with real disposable income.
   b. varies inversely with real disposable income.
   c. changes with changes in wealth.
   d. equals planned saving.

2. The 45-degree reference line
   a. indicates planned expenditures.
   b. is a line along which planned expenditures equal real national income.
   c. is the consumption function.
   d. is the autonomous investment function.

3. If planned investment is autonomous, then
   a. it is independent of real national income.
   b. it is independent of the interest rate.
   c. it varies directly with real national income.
   d. it varies inversely with real national income.

4. At that level of income where the planned expenditures curve intersects the 45-degree reference line (ignoring G and X),
   a. equilibrium exists.
   b. unplanned inventory changes equal zero.
   c. planned saving equals planned investment.
   d. All of the above

5. Changes in autonomous expenditures
   a. affect the 45-degree reference line.
   b. shift the planned expenditures curve.
   c. are movements along the planned expenditures curve.
   d. lead to equal increases in real national income.

6. The consumption function analyzed in the text
   a. has an autonomous component that varies with income.
   b. indicates that real consumption falls as real income rises.
   c. shifts if autonomous consumption changes.
   d. is the same as the autonomous investment function.

7. If total planned expenditures exceed real national income, then
   a. business inventories will rise involuntarily.
   b. business inventories will fall involuntarily.
   c. equilibrium exists.
   d. real national income will fall.

8. If total planned expenditures are less than real national income, then
   a. business inventories will rise involuntarily.
   b. business inventories will fall involuntarily.
   c. equilibrium exists.
   d. real national income will rise.

9. Which of the following is most **UNLIKE** the others?
   a. consumption function
   b. investment function
   c. 45-degree reference line
   d. planned expenditures curve

10. If business inventories rise involuntarily, then
    a. equilibrium exists.
    b. total planned expenditures are less than real national income.
    c. real national income will rise.
    d. businesses will hire more labor.

11. In the short run, total employment and real national income/output
    a. are inversely related.
    b. are directly related.
    c. are independent of each other.
    d. reflect the law of demand.

12. If planned saving exceeds planned investment, then (ignoring government and foreign transactions),
    a. total planned expenditures are less than real national income.
    b. national income exceeds planned expenditures.
    c. business inventories will rise involuntarily, and national income will fall.
    d. All of the above

13. In macroeconomic equilibrium (ignoring government and foreign transactions),
    a. planned saving equals planned investment.
    b. actual inventories equal actual investment.
    c. actual saving does not equal actual investment.
    d. saving plus investment equals consumption.

14. As the MPC rises, the multiplier
    a. falls.
    b. rises.
    c. is unaffected.
    d. changes in an unpredictable way.

15. If the MPC is 1/2, then (ignoring price level effects)
    a. the multiplier is 12.
    b. changes in autonomous income lead to equal changes in national income.
    c. shifts in the planned expenditures curve lead to a change in equilibrium real national income that equals twice the value of the shift.
    d. the APC must fall as real income falls.

16. The multiplier
    a. relates changes in autonomous expenditures to changes in equilibrium real national income.
    b. deals with shifts in the planned expenditures curve.
    c. implies that economic fluctuations are magnified.
    d. All of the above

17. When the price level rises,
    a. the planned expenditures curve shifts downward.
    b. the 45-degree reference line shifts upward.
    c. autonomous expenditures rise.
    d. the multiplier effect is increased.

18. If the planned expenditures curve shifts downward when the price level rises, and upward when the price level falls, then
    a. the multiplier effect is lessened.
    b. the economy can pull out of a recession faster.
    c. economic fluctuations due to shocks will be lessened.
    d. All of the above

19. *Analogy*: Saving is to income as _____ is to wealth.
    a. income
    b. savings
    c. investment
    d. consumption

20. Saving plus consumption equals
    a. investment.
    b. aggregate demand.
    c. disposable income.
    d. 1.

21. Which of the following causes the consumption function to shift?
    a. an increase in real disposable income
    b. a decrease in real disposable income
    c. an increase in wealth
    d. an increase in investment

22. If autonomous consumption is positive, then
    a. the vertical intercept of the consumption function is positive.
    b. the APC falls as real disposable income rises.
    c. dissaving occurs at very low real disposable income levels.
    d. All of the above

23. Autonomous consumption
    a. varies with disposable national income.
    b. varies with wealth.
    c. changes lead to movements along a given consumption curve.
    d. if positive, means that the vertical intercept of the consumption function is zero.

24. If the MPC = 0.8 then the
    a. APC = 0.2.
    b. MPS = 0.2.
    c. APS = 0.2.
    d. vertical intercept of the consumption function is positive.

25. If real disposable income rises by $100 and consumption rises by $75, then
    a. the APC = 0.75.
    b. the MPC = 0.25.
    c. the MPS = 0.25.
    d. the APS = 0.75.

26. If the APC falls as real disposable income rises, then
    a. the APS must rise as real disposable income rises.
    b. the APS is constant.
    c. the MPC must be falling.
    d. the vertical intercept of the saving function must be positive.

27. If the APC falls as real disposable income rises, the
    a. vertical intercept of the consumption function is positive.
    b. vertical intercept of the saving function is negative.
    c. APS rises from a negative number to zero to a positive number.
    d. All of the above

MATCHING

Choose the item in Column (2) that best matches an item in Column (1).

(1)
a. planned investment is less than planned saving
b. equilibrium
c. recession
d. reciprocal of MPS
e. APC
f. MPC
g. autonomous consumption

(2)
h. planned expenditures equal national income
i. falling national income
j. multiplier
k. planned expenditures are less than national income
l. nonincome determinants of consumption
m. total consumption divided by total income
n. change in consumption divided by change in income

WORKING WITH GRAPHS

1. Using the graphs provided, answer the questions that follow.

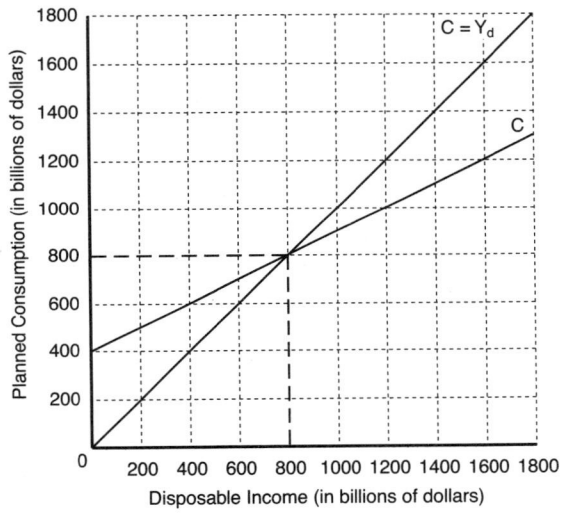

(a)

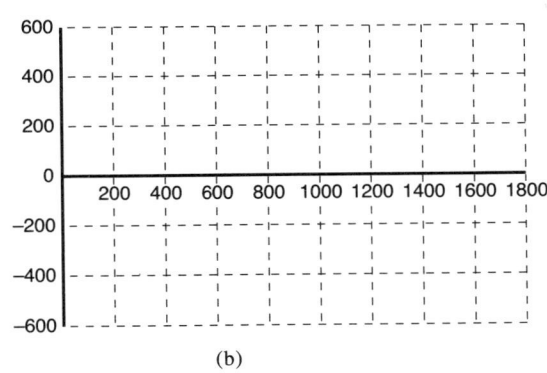

(b)

   a. Graph the saving function in the space provided in panel (b).
   b. What is the break-even level of disposable income?
   c. What is the MPC? What is the MPS?
   d. What is the APC at $800 billion of disposable income?
   e. What is the APC at $1,600 billion of disposable income?
   f. What is the APC at $1,200 billion of disposable income?

2. Given the following information about a hypothetical economy, complete the table below and then represent this economy in a planned expenditures diagram and a saving/investment diagram in the space provided. Include a 45-degree reference line and indicate the level of equilibrium national income. (The figures are given in billions of dollars per year.)

$C = \$200 + 0.8Y \quad I = \$160$

Y	I	C	S	Total Planned Expenditures	Inventory Changes
800	160	840	−40	1000	−200
900	160	920	−20	1080	−180
1500	160	1400	100	1560	−60
1800	160	1640	160	1800	0
2000	160	1800	200	1960	40
2400	160	2120	280	2280	120

where C = planned consumption   S = planned saving
       Y = real national income I = planned investment

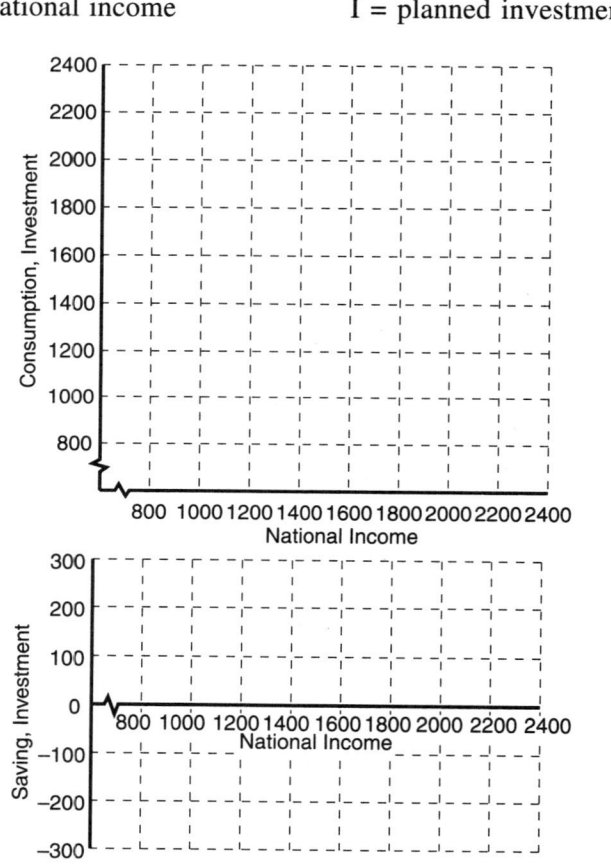

3. Consider the graphs below, then answer the questions that follows. (Note that G = government spending and X = net imports.)

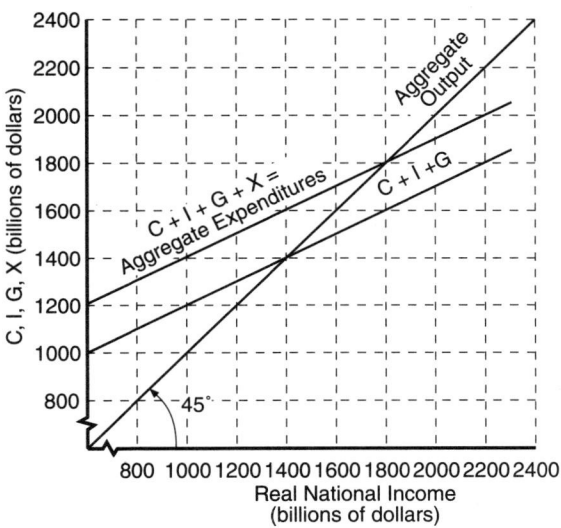

a. What is the value of net exports?
b. What is the MPC for this economy?
c. What is the value of the equilibrium level of real national income?
d. What is the multiplier for this economy?

PROBLEMS

1. Suppose that for a particular economy, MPS = 1/4. Complete the following table under the assumption that autonomous investment has just increased by $2,000 (using the simple Keynesian model).

	Change in Income	Change in Consumption	Change in Saving
Round 1	2,000	_____	_____
Round 2	1,500	_____	_____
Round 3	_____	_____	_____
All other rounds	_____	_____	_____
Total	_____	_____	_____

2. Answer the following, assuming a simple Keynesian economy:
   a. If the MPC = 3/4, and the current equilibrium level of real national income is $11,240 billion, what will be the new equilibrium level of income if autonomous investment falls by $10 billion?
   b. Given the same initial equilibrium level of real national income and the same decrease in investment, what would be the new equilibrium if MPC is 4/5 rather than 3/4?

136 CHAPTER 12: CONSUMPTION, INCOME, AND THE MULTIPLIER

3. Consider the graphs below, then answer the questions that follow. [Ignore G and X]

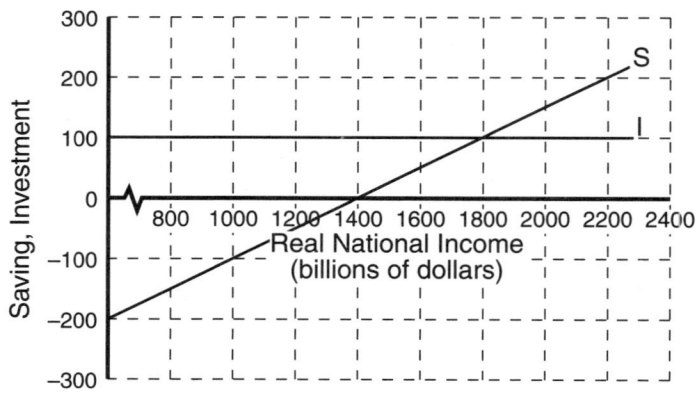

- a. What is the value of autonomous saving?
- b. What is the value of autonomous consumption?
- c. What is the value of autonomous investment?
- d. At a real national income of $2,000 billion, are planned expenditures equal to, greater than, or less than real national income? Why?
- e. At a real national income of $1,600 billion, are unplanned inventories rising, falling, or remaining constant? Why?
- f. What is the equilibrium level of real national income? Why?

4. Assume that you know that the MPC for a particular economy is three-fourths.
   - a. If the break-even income is $10,000, what will be the level of consumption if income is $14,000?
   - b. If income drops to $8,000, what will be the level of consumption? Of saving?

ANSWERS TO CHAPTER 12

COMPLETION QUESTIONS

1. flow; stock
2. disposable income; rises, rises
3. income; 1
4. divided by; MPS
5. independent of; positive; falls
6. exceed
7. expectations, wealth, population
8. inversely; profit expectations, innovation and technology, business taxes
9. national income; equilibrium; disequilibrium
10. planned expenditures
11. fall; increase; rise; decrease
12. upward; rise
13. negative; fall; greater than

TRUE-FALSE QUESTIONS

1. F The APC plus the APS equals 1.0.
2. T
3. T
4. T
5. F No shift, just a movement along (up) the C-function.
6. T
7. F At every point along the 45-degree line, planned expenditures equal national income.
8. F Autonomous here means independent of income.
9. T
10. T
11. T
12. F Planned expenditures would exceed national income.
13. T
14. T
15. T
16. T

MULTIPLE CHOICE QUESTIONS

1.c; 2.b; 3.a; 4.d; 5.b; 6.c; 7.b; 8.a; 9.c; 10.b;
11.b; 12.d; 13.a; 14.b; 15.c; 16.d; 17.a; 18.d; 19.b; 20.c;
21.c; 22.d; 23.b; 24.b; 25.c; 26.a; 27.d.

MATCHING

a and k; b and h; c and i; d and j; e and m; f and n; g and l

WORKING WITH GRAPHS

1. a.

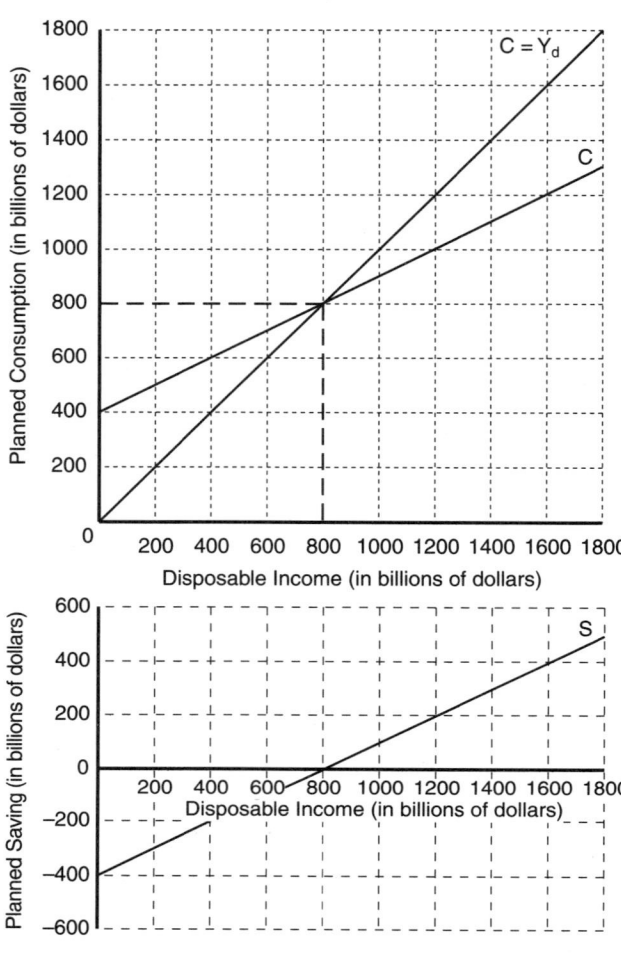

b. $800 billion ; c. 1/2; 1/2 d. 1 e. 3/4 f. 5/6

2.

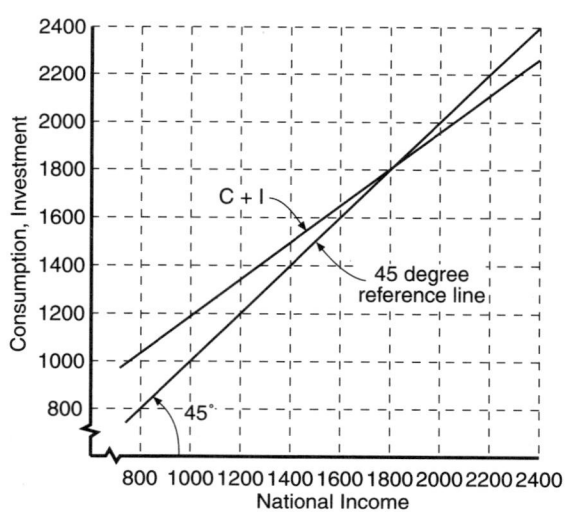

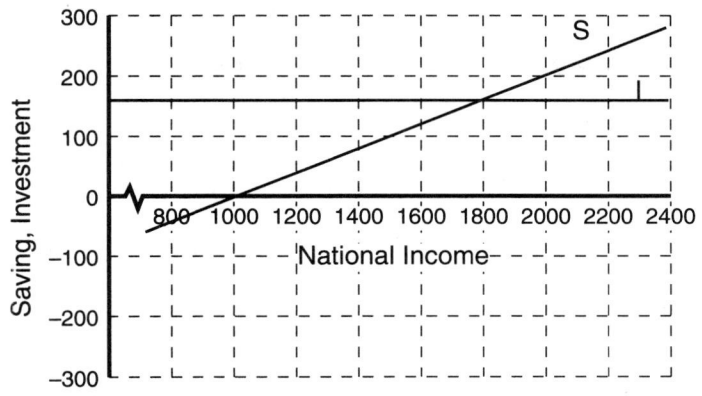

Y	I	C	S	Total Planned Expenditures	Inventory Changes
800	160	840	-40	1000	-200
900	160	920	-20	1080	-180
1500	160	1400	100	1560	-60
1800	160	1640	160	1800	0
2000	160	1800	200	1960	40
2400	160	2120	280	2280	120

3. a. $200 billion; b. 0.5; c. $1800 billion; d. 2.0

PROBLEMS

1.

	Change in Income	Change in Consumption	Change in Saving
Round 1	2,000	$1,500.00	$ 500.00
Round 2	1,500	1,125.00	375.00
Round 3	1,125	843.75	281.25
All other rounds	3,375	2,531.25	843.75
Total	$8,000	$6,000.00	$2,000.00

2. a. $11,200 billion
   b. $11,190 billion

3. a. - $200 billion
   b. $200 billion
   c. $100 billion
   d. Less than, because planned saving > planned investment.
   e. Falling, because at that income level planned investment > planned saving.
   f. $1,800 billion, because only here does planned saving equal planned investment.

4. a. $13,000
   b. $8,500; -$500 (dissaving)

GLOSSARY TO CHAPTER 12

Autonomous consumption That part of consumption that is independent of (does not depend on) the level of disposable income. Changes in autonomous consumption shift the consumption function.

Average propensity to consume (APC) Consumption divided by disposable income; for any given level of income, the proportion of total disposable income that is consumed.

Average propensity to save (APS) Saving divided by disposable income; the proportion of total disposable income that is saved.

Capital goods Producer durables; nonconsumable goods that firms use to make other goods.

Consumption Spending on new goods and services out of a household's current income. Whatever is not consumed is saved. Consumption includes such things as buying food and going to a concert.

Consumption function The relationship between the amount consumed and disposable income. A consumption function tells us how much people plan to consume at various levels of disposable income.

Consumption goods Goods bought by households to use up, such as food and movies.

Dissaving Negative saving; a situation in which spending exceeds income. Dissaving can occur when a household is able to borrow or use up existing assets.

45-degree reference line The line along which planned real expenditures equal real national income per year.

Investment Spending by businesses on things such as machines and buildings, which can be used to produce goods and services in the future. The investment part of total output is the portion that will be used in the process of producing goods in the future.

Lump-sum tax A tax that does not depend on income or the circumstances of the taxpayer. An example is a $1,000 tax that every family must pay, irrespective of its economic situation.

Marginal propensity to consume (MPC) The ratio of the change in consumption to the change in disposable income. A marginal propensity to consume of 0.8 tells us that an additional $100 in take-home pay will lead to an additional $80 consumed.

Marginal propensity to save (MPS) The ratio of the change in saving to the change in disposable income. A marginal propensity to save of 0.2 indicates that out of an additional $100 in take-home pay, $20 will be saved. Whatever is not saved is consumed. The marginal propensity to save plus the marginal propensity to consume must always equal 1, by definition.

Multiplier The ratio of the change in the equilibrium level of real national income to the change in autonomous expenditures; the number by which a change in autonomous investment or autonomous consumption, for example, is multiplied to get the change in the equilibrium level of real national income.

Saving The act of not consuming all of one's current income. Whatever is not consumed out of spendable income is, by definition, saved. Saving is an action measured over time (a flow), whereas *savings* are an existing accumulation of wealth resulting from the act of saving in the past.

Wealth The stock of assets owned by a person, household, firm, or nation. For a household, wealth can consist of a house, cars, personal belongings, stocks, bonds, bank accounts, and cash.

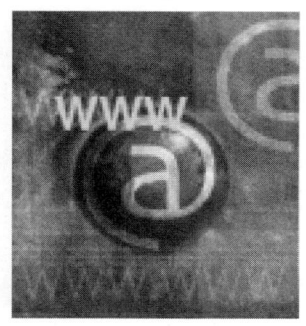

CHAPTER 13
FISCAL POLICY

LEARNING OBJECTIVES

After you have studied this chapter, you should be able to

1. define fiscal policy, direct expenditure offsets, automatic or built-in stabilizers, crowding out, recognition time lag, action time lag, effect time lag, Ricardian equivalence theorem, and supply-side economics;

2. recognize the proper fiscal policy required to eliminate recessionary gaps and inflationary gaps;

3. distinguish between the effects of fiscal policy when the economy is operating on the LRAS curve and when it is not;

4. recognize how direct and indirect offsets limit the effectiveness of fiscal policy;

5. indicate how an expansionary fiscal policy can cause net exports to fall;

6. distinguish between discretionary fiscal policy and automatic fiscal policy;

7. enumerate the major problems associated with conducting fiscal policy;

8. recognize how changes in marginal tax rates can have supply-side effects;

9. distinguish among the three fiscal policy time lags.

CHAPTER OUTLINE

1. Fiscal policy is the discretionary changing of government expenditures and/or taxes in an attempt to achieve such national economic goals as high employment and price stability.
   a. If a recessionary gap exists, then expansionary fiscal policy is in order; if government expenditures increase (or lump-sum taxes fall) the aggregate demand curve shifts rightward and the recessionary gap can be eliminated, at a higher price level.
   b. If an inflationary gap exists, then contractionary fiscal policy is called for; if government expenditures decrease (or lump-sum taxes increase) the aggregate demand curve shifts leftward and the inflationary gap can be eliminated, at a lower price level.
   c. If the economy is already operating on the LRAS curve, shifts in the AD curve lead to *temporary* increases (decreases) in real GDP which are untenable because they are off the

LRAS curve; in the long run, input owners revise their expectations upward (downward) and the SRAS curve shifts upward (downward). In the long run, GDP will be at the level along the LRAS curve, and the price level change will be greater than the change in the short run.

2. There are various factors that offset fiscal policy and thereby limit its effectiveness.
   a. Indirect offsets to fiscal policy:
      i. If government expenditures are financed by borrowing (deficit spending), then the interest rate may rise, which will cause a reduction in (a) business investment, and (b) household expenditures on such durable goods as housing and automobiles.
      ii. If households perceive deficit spending as an increase in their future tax liabilities, the Ricardian equivalence theorem predicts that they will save more, and hence the AD curve may not shift at all: household current consumption falls by the amount that G rises.
   b. Direct fiscal offsets arise when government expenditures compete with the private sector, so that increases in government spending are offset by decreases in private investment.
   c. Supply-side effects can result from fiscal policy effects of changing tax rates: changes in marginal tax rates can affect the choice between leisure and labor, thereby affecting how much people work.

3. The recognition, action, and effect time lags reduce the effectiveness of fiscal policy.

4. A progressive income tax and unemployment compensation are two examples of automatic, or built-in, stabilizers; they are not discretionary, and they move the economy automatically toward high employment levels.

5. During normal times when there is not excessive unemployment or inflation, fiscal policy actions by the Congress have proven to be relatively ineffective—usually too little too late to help in minor recessions.

6. Each year that the government spends more than it collects in revenues, it runs a deficit; the accumulation of deficits results in government debt.
   a. A government budget deficit occurs over time and thus is a flow; government debt is an accumulated stock measured at a point in time.
   b. All public indebtedness taken together is the gross public debt; the net public debt is gross public debt minus all government interagency borrowing.

KEY TERMS
Recognition time lag
Action time lag
Effect time lag

Supply-side economics
Gross public debt
Net public debt

KEY CONCEPTS
Fiscal policy
Ricardian equivalence theorem
Automatic, or built-in, stabilizers

Crowding-out effect
Direct expenditure offsets
Indirect expenditure offsets

COMPLETION QUESTIONS
Fill in the blank, or circle the correct term.

1. Discretionary fiscal policy is defined as a(n) _____ change in taxes and/or government spending in order to change equilibrium national income and employment.

2. If a recessionary gap exists, it can be offset by (contractionary, expansionary) fiscal policy; such a policy entails (decreasing, increasing) government expenditures or (decreasing, increasing) taxes, which will cause the aggregate demand curve to shift (leftward, rightward); national output should (fall, rise) and the price level should (fall, rise).

3. If an inflationary gap exists, it can be eliminated if government expenditures (decrease, increase) or if taxes are (decreased, increased); this will cause the AD curve to shift (leftward, rightward) and national income will (fall, rise).

4. If the economy is *already* operating on its long-run aggregate supply curve, then fiscal policy actions which shift the AD curve will cause real national income to change (temporarily, permanently) and the price level will change (more, less) in the long run, relative to the short run.

5. If government expenditures are financed by borrowing, a federal budget (deficit, surplus) will result, which may cause the interest rate to (fall, rise), which in turn will cause business investment and household consumption on (durable, nondurable) goods to (fall, rise); hence fiscal policy effects will be (reduced, increased).

6. If households perceive government deficit spending as an increase in their future tax liabilities, they may save (less, more) according to the _____ theorem; hence fiscal policy effects will be (reduced, increased).

7. To the extent that government expenditures compete with the private sector, then such expenditures will (induce more, discourage) business investment expenditures; hence fiscal policy effects of an increase in government spending will be (reduced, enlarged).

8. If U.S. government deficit spending causes market interest rates to rise, businesses will want to (increase, decrease) investment spending, and households will desire to (increase, decrease) spending on durable goods; hence the expansionary effects of an increase in government spending will be (reduced, enlarged) by this indirect expenditure offset.

9. Supply-side effects can result from fiscal policy tax changes; if marginal tax rates rise this can induce laborers to substitute _____ for _____.

10. A nation that attempts to generate a net increase in its tax revenues by (reducing; raising) marginal income tax rates to attract labor and capital from other nations engages in _____.

11. If government expenditures or taxes change over the business cycle without deliberate action taken by Congress, this is referred to as automatic fiscal policy, or built-in _____; examples of automatic fiscal policy include _____ and _____; automatic fiscal policy (increases, decreases) the magnitude of business cycle fluctuations.

12. Discretionary fiscal policy is (easy, difficult) to conduct because it usually takes (little, much) time for Congress to enact such policy.

13. If the public perceives that deficit spending creates future tax liabilities, and if people wish to leave money to their heirs, then current saving may well (decrease, increase); hence the net effect of deficit spending on interest rates is (to lower them, to raise them, uncertain).

14. There are three time lags that hamper fiscal policy: _____, _____, and _____. The existence of time lags makes conducting fiscal policy (easier, harder) for policymakers.

15. Because of automatic stabilizers, when the economy is in an expansion phase government transfers (rise, fall) and tax revenues (rise, fall); hence expansions (other things constant) generate government budget (surpluses, deficits).

16. The existence of automatic stabilizers makes our economy (less, more) stable; they also make it (difficult, easy) to distinguish discretionary from automatic fiscal policy.

17. The gross public debt (includes, subtracts out) all government interagency borrowing, whereas the net public debt (includes, subtracts out) interagency borrowing within the government sector.

TRUE-FALSE QUESTIONS
Circle the **T** if the statement is true, the **F** if it is false. Explain to yourself why a statement is false.

T F   1. Fiscal policy may involve changes in taxes and/or government spending.

T F   2. If an inflationary gap exists, fiscal policy calls for increased government spending and/or reduced taxes.

T F   3. If a recessionary gap exists, proper fiscal policy requires a federal government budget surplus—or a larger surplus if one already exists.

T F   4. If an economy is already operating on its LRAS curve, an expansionary fiscal policy will, eventually, cause the price level to rise by less than it would if the economy had been operating at a SRAS curve.

T F   5. If government expenditures are financed by borrowing, a federal deficit is created which could cause interest rates to rise.

T F   6. If interest rates rise as a result of deficit spending, expansionary fiscal policy effects will be magnified.

T F   7. If interest rates rise as a result of deficit spending, then businesses and households may choose to cut back on purchases of investment goods and durable goods.

T F   8. If households perceive an increase in federal deficit spending as an increase in their future tax liabilities they may save more now, which would reduce the effects of expansionary fiscal policy.

T F   9. If government expenditures directly compete with the output of the private sector, then business investment will fall and tend to offset the effects of such a fiscal policy.

T F   10. Crowding out implies that if federal deficits cause interest rates to rise, businesses will reduce investments and this will tend to offset fiscal policy effects.

T F   11. Because of the time lags involved in fiscal policy, policymakers can more easily achieve national economic goals, because they have more time to solve the problem.

T F   12. If federal deficit spending causes interest rates to rise, households will purchase more consumer durables and businesses will invest more.

T F   13. If fiscal policy is pursued by raising marginal tax rates, laborers may choose to work less and businesses might choose to make fewer investments.

T F 14. A federal government debt is a flow over time, whereas the federal government's deficit is a stock measured at a point in time.

MULTIPLE CHOICE QUESTIONS
Circle the letter that corresponds to the best answer.

1. Discretionary fiscal policy
   a. deals with automatic stabilizers.
   b. is relatively easy to conduct.
   c. deals with actions by Congress and the president intended to affect economic performance.
   d. calls for stabilizing changes in the money supply via changes in government spending or taxes.

2. If a recessionary gap exists, proper fiscal policy could entail
   a. increased government spending.
   b. decreased taxes.
   c. deficit spending.
   d. All of the above

3. If government expenditures rise to counteract a recessionary gap,
   a. the AD curve shifts rightward.
   b. the AD curve shifts leftward.
   c. taxes must rise to finance such expenditures.
   d. the price level will fall.

4. If an inflationary gap exists,
   a. contractionary fiscal policy is called for.
   b. government spending should rise to cause the price level to fall.
   c. expansionary fiscal policy is called for.
   d. taxes should fall to stimulate the economy.

5. If a recessionary gap exists, then
   a. the equilibrium income exceeds the full employment income level.
   b. it can be filled by increases in government spending or decreases in taxes, or some combination of both.
   c. it can be filled by decreases in government spending or increases in taxes, or some combination of both.
   d. the economy is in an expansion phase.

6. If the economy is operating on its short-run aggregate supply curve and an inflationary gap exists,
   a. a contractionary fiscal policy is called for.
   b. leftward shifts in AD that cause some unemployment will be helpful.
   c. a contractionary fiscal policy will eventually change only the price level.
   d. All of the above

7. If an inflationary gap exists, it can most efficiently be eliminated by some combination of
   a. increases in government spending and decreases in taxes.
   b. decreases in government spending and decreases in taxes.
   c. decreases in government spending and increases in taxes.
   d. increases in government spending and increases in taxes.

8. Which of the following will **NOT** offset fiscal policy?
   a. the multiplier effect
   b. government spending financed by increased taxes
   c. government spending financed by borrowing (deficit spending)
   d. automatic stabilizers

9. Which of the following can offset an expansionary fiscal policy?
   a. higher interest rates resulting from deficit spending
   b. private investment falling in areas competing with government expenditures
   c. perceptions by households that larger deficits imply increased future tax liabilities
   d. All of the above

10. Which of the following may well result from increased government borrowing resulting from deficit spending?
    a. increased purchases of consumer durable goods
    b. increased business investment
    c. higher interest rates
    d. All of the above

11. If taxes fall, then
    a. the planned expenditures curve shifts downward.
    b. the aggregate demand curve shifts to the right.
    c. the aggregate supply curve shifts to the left.
    d. national income will fall.

12. If government expenditures exceed tax receipts then, other things being constant,
    a. a surplus exists.
    b. a balanced budget exists.
    c. a deficit exists.
    d. the economy must contract.

13. If marginal tax rates rise,
    a. laborers may choose less income (work), which is taxed, and more leisure, which is not taxed.
    b. the tax base could shrink.
    c. productivity could fall eventually, as business investment falls.
    d. All of the above

14. Fiscal policy
    a. if discretionary, only involves changing taxes.
    b. is difficult to implement because of time lag problems.
    c. if automatic, is destabilizing.
    d. All of the above

15. Choose the statement that is **NOT** true.
    a. Time lags make fiscal policy difficult.
    b. Fiscal policy is conducted solely by the executive branch of the U.S. government.
    c. Crowding out reduces the impact of an expansionary fiscal policy.
    d. The Ricardian equivalence theorem implies that fiscal policy may be quite ineffective.

MATCHING

Choose an item in Column (2) that best matches an item in Column (1).

(1)
a. stabilization policy
b. discretionary fiscal policy
c. automatic stabilizer
d. recessionary gap
e. inflationary gap
f. time lags
g. deficit spending

(2)
h. recognition, action, effect
i. change in tax law
j. tax receipts less than government spending
k. unemployment compensation
l. recession period
m. inflationary period
n. conscious attempt to achieve high employment and price stability

WORKING WITH GRAPHS

1. Consider the graphs below, then answer the questions that follow.

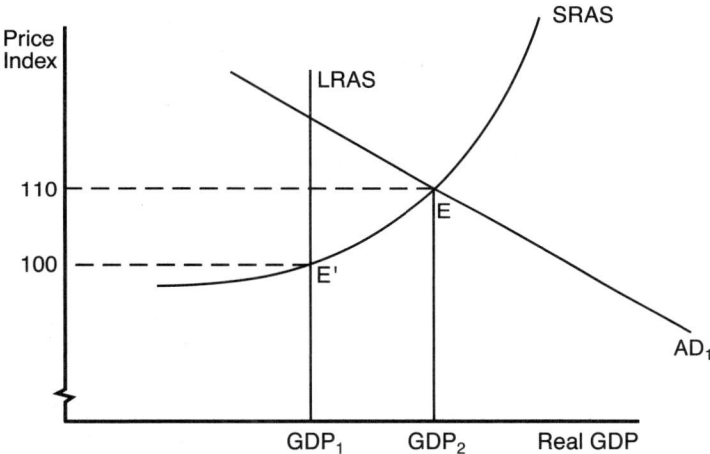

a. What is the short-run equilibrium level of real GDP? What type of gap exists? Is this real income level sustainable?
b. What type of fiscal policy would you recommend? Be specific.
c. Under your fiscal policy, through what point will the new AD curve shift?
d. The long-run result of your fiscal policy is to cause what to happen to real GDP? To the price index?

2. Consider the graphs below, then answer the questions that follow.

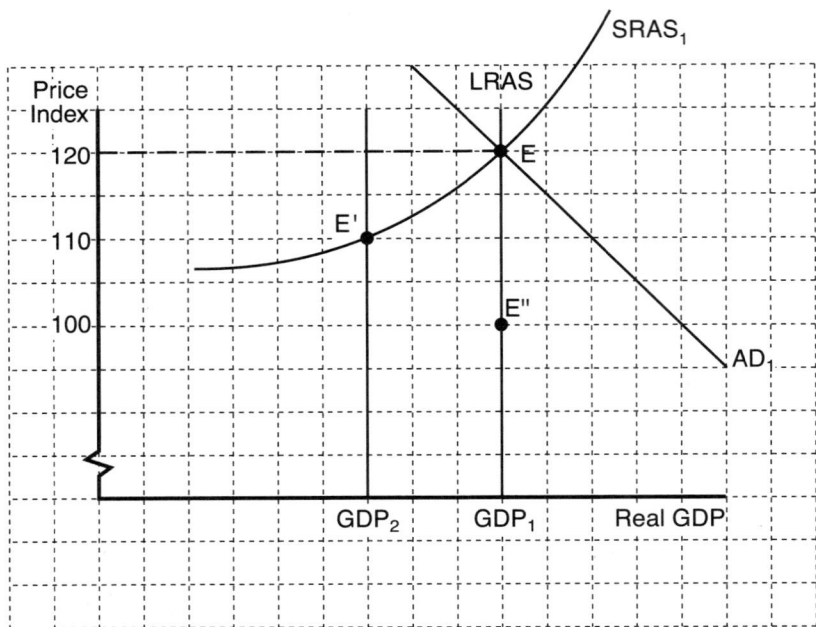

a. What is the short-run equilibrium level of real GDP? The long-run equilibrium level of real GDP? The short-run equilibrium price level?
b. Assume that it is desirable to get the short-run equilibrium price level to 110. What type of fiscal policy would you suggest? (Be specific.) (Through what point will the new AD curve come?) What will be the new short-run level of real GDP? Is this level sustainable? Why not?
c. Continuing (b) above, what will happen to the short-run aggregate supply curve? Why?
d. What will be the long-run level of real GDP?

ANSWERS TO CHAPTER 13

COMPLETION QUESTIONS

1. deliberate, or conscious
2. expansionary; increasing; decreasing; rightward; rise, rise
3. decrease; increased; leftward; fall
4. temporarily; more
5. deficit; rise; durable; fall; reduced
6. more; Ricardian equivalence; reduced
7. discourage; reduced
8. decrease; decrease; reduced
9. leisure; income resulting from working
10. reducing; international tax competition
11. stabilizers; progressive tax structure; unemployment compensation; decreases
12. difficult; much
13. increase; uncertain
14. recognition, action, effect; harder
15. fall; rise; surpluses
16. more; difficult
17. includes; subtracts out

TRUE-FALSE QUESTIONS

1. T
2. F   An inflationary gap calls for a decrease in government spending and/or an increase in taxes.
3. F   A recessionary gap calls for deficit spending.
4. F   No, the price level will change by more because output won't change.
5. T
6. F   No, higher interest rates will cause offsetting expenditure reductions in the private sector.
7. T
8. T
9. T
10. T
11. F   Time lags make fiscal policy more difficult because of the uncertainty they generate.
12. F   No, less of such expenditures will occur in response to a higher interest rate.
13. T
14. F   A federal government deficit is a flow, whereas the government's debt is a stock.

MULTIPLE CHOICE QUESTIONS

1.c;   2.d;   3.a;   4.a;   5.b;   6.d;   7.c;   8.a;   9.d;   10.c;
11.b;   12.c;   13.d;   14.b;   15.b.

MATCHING

a and n; b and i; c and k; d and l; e and m; f and h; g and j

WORKING WITH GRAPHS

1. a. GDP_2; inflationary gap; no, because it is above the full employment level of real GDP
   b.   Contractionary; Reduce government spending and/or increase taxes.
   c.   E'
   d.   Fall to GDP_1; fall to 100

2. a. GDP_1; GDP_1; 120.
   b. Contractionary; reduce government spending, increase taxes; E'; GDP_2; No, because it is below the real GDP consistent with LRAS.
   c. It will shift downward (rightward) as factors of production become accustomed to the lower price level.
   d. GDP_1

GLOSSARY TO CHAPTER 13

Action time lag The time required between recognizing an economic problem and implementing policy to solve it. The action time lag is quite long for fiscal policy, which requires congressional approval.

Automatic, or **built-in, stabilizers** Special provisions of certain federal programs that cause changes in desired aggregate expenditures without the action of Congress and the president. Examples are the federal tax system and unemployment compensation.

Crowding-out effect The tendency of expansionary fiscal policy to cause a decrease in planned investment or planned consumption in the private sector; this decrease results from the rise in interest rates.

Direct expenditure offsets Actions taken by the private sector that offset government fiscal policy actions. Any increase in government spending in an area that competes with the private sector will have some direct expenditure offset.

Effect time lag The time that elapses between the implementation of a policy and the results of that policy.

Fiscal policy The discretionary changing of government expenditures or taxes to achieve national economic goals, such as high employment with price stability.

Gross public debt All federal government debt irrespective of who owns it.

Net public debt Gross public debt minus all government interagency borrowing.

Recognition time lag The time required to gather information about the current state of the economy.

Ricardian equivalence theorem The proposition that an increase in the government budget deficit has no effect on aggregate demand.

Supply-side economics The notion that creating incentives for individuals and firms to increase productivity will cause the aggregate supply curve to shift outward.

CHAPTER 14
MONEY, BANKING, AND CENTRAL BANKING

LEARNING OBJECTIVES

After you have studied this chapter, you should be able to

1. list and explain the four functions of money;

2. explain why people typically prefer to use money rather than engaging in barter;

3. define liquidity and rank assets according to their liquidity;

4. compare alternative monetary standards and explain the key aspects of a fiduciary monetary system;

5. distinguish between the transactions approach and the liquidity approach to measuring money;

6. list the components of M1 and explain the relationship between M1 and M2;

7. provide key rationales for the existence of financial intermediaries such as banks and explain why financial intermediation takes place across national borders;

8. identify the three fundamental duties of most of the world's central banks and enumerate the specific functions of the U.S. Federal Reserve System;

9. discuss the basic structure of the Federal Reserve System.

CHAPTER OUTLINE

1. There are four traditional functions of money.
   a. Money is a medium of exchange: money is that for which people exchange their productive services or that which they give for goods and services.
   b. Money is a unit of accounting: the monetary unit is used to value goods and services relative to each other.
   c. Money acts as a store of value: money is an asset that is a convenient store of generalized purchasing power.
   d. Money is a standard of deferred payment: money is used to pay future obligations or debts.

2. Liquidity is the degree to which an asset can be acquired or disposed of without loss in terms of nominal value and with small transactions costs.
   a. Money is the most liquid of all assets.
   b. Different goods have served as money throughout history.

3. Today the United States is on a fiduciary monetary system.
   a. The dollar is money in the United States because it is acceptable to virtually everyone in exchange for goods and services.
   b. Another reason the dollar is money in the United States is because it has predictability of value in the future.

4. There are two basic approaches to measuring the money supply.
   a. The transactions approach to measuring money—which emphasizes the M1 measure of money—stresses that the essence of money is that it is a medium of exchange.
      i. M1 includes currency—monetary coins and paper money.
      ii. M1 also includes checkable deposits (accounts on which people can write checks).
      iii. M1 also includes the value of traveler's checks issued by nonbank institutions.
   b. The liquidity approach to measuring money—which emphasizes the M2 measure of money—stresses that money is a highly liquid asset; such assets have an unchanging nominal value.
      i. M2 includes all the items in M1.
      ii. M2 also includes savings deposits in all depository institutions.
      iii. M2 time deposits, money market deposit accounts, and retail money market mutual fund balances.

5. Financial intermediaries, such as banks, make indirect finance possible; their function is to channel funds from ultimate lenders to ultimate borrowers.
   a. One reason that savers may use the services of a financial intermediary instead of lending funds directly is that they may face a problem of asymmetric information: Prospective borrowers may have better knowledge of their own current and future prospects than do potential lenders.
      i. Adverse selection is the possibility that prospective borrowers desire to borrow funds to use in unworthy projects; one reason that financial intermediaries exist is that they specialize in evaluating the creditworthiness of prospective borrowers.
      ii. Moral hazard is the possibility that a borrower may engage in riskier behavior after receiving a loan; another reason that financial intermediaries exist is to monitor the ongoing performance of borrowers.
   b. Financial intermediaries also make it possible for many people to pool their funds together to take advantage of lower funds management costs that can result from the increased size, or scale, of savings managed by a single institution.
   c. Every financial intermediary has its own sources of funds, which are its liabilities; it also has its own uses of assets.
   d. In the absence of capital controls that restrict movements of funds across borders, people may wish to engage in international financial diversification by engaging in direct or indirect finance across national borders; today an increasing number of financial intermediaries, including banks, take part in the process of international financial intermediation.

6. Banking structures of the world's nations have features distinctive to each country, but nations' central banks tend to perform similar functions.
   a. In some countries, such as Germany, banks traditionally are the predominant sources of finance for businesses, and a few banks tend to dominate; in others, such as the United States, bank finance is a much smaller portion of total direct and indirect finance, and there are many banks of various sizes.

b. Central banks in most countries tend to have three essential duties.
   i. Central banks perform banking functions for national governments.
   ii. Central banks provide financial services for private banks.
   iii. Central banks conduct monetary policies.

7. The Federal Reserve System, the central bank of the United States, was established in 1913 to counter the periodic financial panics that had occurred.
   a. The Fed organizational chart shows a Board of Governors, 12 Federal Reserve district banks having 25 branches, and a Federal Open Market Committee that determines Fed policy actions.
   b. Commercial banks and other depository institutions are required by the Fed to keep a certain percentage of their deposits on reserve with Fed district banks.

8. There are eight major functions performed by the Fed.
   a. It supplies the country with fiduciary currency.
   b. It provides a system of check collection and clearing.
   c. It holds depository institutions' reserves.
   d. It acts as the government's fiscal agent.
   e. It supervises depository institutions.
   f. It acts as a lender of last resort.
   g. It regulates the money supply, which is its most important function.
   h. It intervenes in foreign currency markets.

KEY TERMS
Money market deposit accounts
Commercial bank
M1
M2
Asymmetric information
Capital controls
Depository institutions
Checkable deposits
Savings deposits

Money supply
Central bank
The Fed
Financial intermediaries
Adverse selection
International financial diversification
Transactions accounts
Certificate of deposit
Money market mutual funds

Traveler's checks
Thrift institution
Assets
Liabilities
Moral hazard
Time deposit
World index fund
Universal banking

KEY CONCEPTS
Liquidity
Fiduciary monetary system
Money
Near monies
Financial intermediation

Liquidity approach
Barter
Transactions approach
Medium of exchange

Unit of accounting
Store of value
Standard of deferred
   payment

COMPLETION QUESTIONS
Fill in the blank, or circle the correct term.

1. Because money is accepted for goods and services, it is used as a(n) _____;
   it also performs the functions of _____, _____,
   and a(n) _____.

2. _____ is the most liquid of all assets, because it maintains its (nominal, real) value.

3. For an exchange to take place in a barter economy, a(n) _____ coincidence of wants must exist; when a money system replaces a barter economy, specialization (decreases, increases) transaction costs.

4. The opportunity cost of holding money is foregone _____ earnings; the benefit to holding money is increased _____.

5. The United States is on a(n) _____ monetary system, which means the dollar is backed by (gold, faith that it can be exchanged for goods); items are used as money because of their _____ and _____ of value.

6. M1 is the _____ approach to measuring money; M1 includes _____, _____, and _____; each can be used to make _____.

7. M2 is the _____ approach to measuring money; M2 components are characterized by high _____ because they maintain their nominal value.

8. The U.S. central bank is the _____; it was established in 1913 to counter the financial _____ that occurred periodically.

9. Depository institutions include commercial banks, _____, _____, and _____; depository institutions are required to keep a certain percentage of their _____ on reserve with Federal Reserve district banks.

10. The Fed has eight major functions: it _____, _____, _____, _____, _____, _____, _____, and _____. The Fed's most important function is _____.

11. When inflation occurs, the price level (rises, falls), and the value of a unit of money (rises, falls).

12. Financial intermediaries perform the function of transferring household _____ to business _____.

13. An asymmetric-information problem of _____ arises when a borrower uses the proceeds from a loan for riskier projects than the lender had anticipated.

14. There are (5, 7, 12) Federal Reserve district banks in the United States.

TRUE-FALSE QUESTIONS
Circle the **T** if the statement is true, the **F** if it is false. Explain to yourself why a statement is false.

T F 1. Exchange in a money economy requires a double coincidence of wants.

T F 2. In a money economy, specialization is encouraged and transaction costs fall, relative to a barter economy.

T F 3. An asset is liquid if it can be disposed of at a low transaction cost without loss of nominal value.

T F 4. There is no opportunity cost to holding money, because it is the most liquid of all assets.

T F 5. M1 and M2 are the same thing.

T F 6. Barter economies are more efficient than money economies.

T F 7. In the United States the dollar is backed by gold and silver.

T F 8. Currency and transactions accounts are money because of their acceptability and their predictability of value.

T F 9. Barter transactions cannot be arranged on the Internet, because Internet transactions require electronic means of payment.

T F 10. The components of M1 are less liquid than the components of M2.

T F 11. The components of M2 all are used as a medium of exchange.

T F 12. The value of M2 always exceeds the value of M1.

T F 13. The Fed requires depository institutions to hold a certain percentage of their deposits on reserve.

T F 14. Currency is the highest percentage of M1.

T F 15. The Fed's most important function is to supply the economy with fiduciary currency.

T F 16. The Fed is prohibited from being a lender of last resort.

T F 17. When the price level falls, the value of money rises.

T F 18. Financial intermediation is the process of transforming business investments into household saving.

T F 19. The potential for a loan applicant (who has not yet received a loan) to have in mind using borrowed funds for riskier projects than she states in her loan application is an example of the moral hazard problem.

MULTIPLE CHOICE QUESTIONS
Circle the letter that corresponds to the best answer.

1. Which of the following is a function of money?
   a. medium of exchange
   b. unit of accounting
   c. store of value
   d. All of the above

2. Which is **NOT** considered money?
   a. checkable deposits
   b. traveler's checks issued by nonbanks
   c. credit cards
   d. currency

3. Which of the following is a characteristic of M2?
   a. high liquidity
   b. medium of exchange
   c. significant changes in nominal value
   d. real value does not vary during inflationary periods

4. Which of the following is an advantage of money over barter?
   a. permits more specialization of labor
   b. does not require a double coincidence of wants
   c. reduces transaction and storage costs
   d. All of the above

5. Which of the following assets are probably the least liquid?
   a. savings account balances
   b. shares of stock in a major corporation
   c. office buildings
   d. automobiles

6. Which of the following is most **UNLIKE** the others?
   a. currency
   b. passbook savings account
   c. checking accounts
   d. traveler's checks issued by nonbanks

7. Which of the following is **NOT** a component of M1?
   a. savings deposits
   b. nonbank traveler's checks
   c. currency
   d. All of the above

8. Which of the following is true regarding U.S. financial institutions?
   a. They are becoming less similar.
   b. The distinctions among them are becoming blurred.
   c. Universal banking is completely prohibited in the United States.
   d. They all have pretty much the same composition of assets and liabilities.

9. Which of the following is **NOT** a part of M1?
   a. nonbank traveler's checks
   b. credit card limits
   c. currency
   d. checkable deposits

10. Which of the following is a depository institution?
    a. commercial bank
    b. savings and loan association
    c. credit union
    d. All of the above

11. *Analogy:* M1 is to the transactions approach as M2 is to the
    a. unit of accounting approach.
    b. liquidity approach.
    c. barter approach.
    d. medium of exchange approach.

12. Which of the following is the most important function of the Fed?
    a. It regulates the money supply.
    b. It supervises depository institutions.
    c. It supplies the economy with fiduciary currency.
    d. It holds depository institutions' reserves.

13. Barter
    a. increases specialization of labor.
    b. makes money less useful.
    c. increases transaction costs.
    d. eliminates the need for a double coincidence of wants.

14. Near monies
    a. are used as a medium of exchange.
    b. are highly liquid.
    c. include currency and demand deposits.
    d. All of the above

15. Under a fiduciary monetary standard, money is backed by
    a. gold.
    b. public faith that money can be exchanged for goods and services.
    c. precious metals that can be exchanged for goods and services.
    d. the Treasury's assets.

16. Currency
    a. is not a component of M2.
    b. earns no interest.
    c. is illiquid.
    d. cannot be used in transactions.

17. Which one of the following is **NOT** a fundamental function that a nation's central bank normally performs?
    a. conducting monetary policy within the nation
    b. making loans to the nation's largest businesses
    c. providing banking services to the nation's government
    d. providing financial services to private banks throughout the nation

18. Which of the following is an example of a moral hazard problem?
    a. After receiving a loan originally intended to upgrade her company's equipment, the owner of an established Internet service provider decides instead to make a risky investment in a new e-commerce company that is entering an already crowded market.
    b. Two young entrepreneurs, who have developed new Internet browser software that they know has some fundamental flaws, nevertheless apply for a loan from a bank so that they can try to market their software to the public.
    c. Three young women apply for a loan with the stated purpose of starting a small business that would sell established software products, when in fact they plan to use the funds to take a chance on selling untested software.
    d. A brilliant computer programmer who has notoriously poor business judgment seeks to convince an investment bank to purchase and market stock in a start-up e-commerce company he is forming.

MATCHING

Choose the item in Column (2) that best matches an item in Column (1).

(1)
a. M1
b. M2
c. currency
d. bank liability
e. adverse selection
f. liquidity

(2)
g. ease of conversion into money
h. checkable deposit
i. coins and paper money
j. transactions approach
k. liquidity approach
l. asymmetric information

PROBLEMS

1. Below you are given hypothetical figures in billions of dollars for various items relating to the supply of money.

Currency	646.6
Checkable deposits	617.0
Retail money market mutual fund shares	905.1
Large-denomination time deposits	2,241.3
Small-denomination time deposits	675.2
Traveler's checks	8.6
U.S. government securities	3,415.2
Savings deposits	3,113.7
Money market deposit accounts	889.3

   a. Find the value of M1.
   b. Find the value of M2.

ANSWERS TO CHAPTER 14

COMPLETION QUESTIONS

1. medium of exchange; unit of accounting, store of value, standard of deferred payment
2. money; nominal
3. double; decreases
4. interest; liquidity
5. fiduciary; faith that it can be exchanged for goods; acceptability, predictability
6. transactions; currency, checkable deposits, traveler's checks; transactions
7. liquidity; liquidity
8. Fed; panics
9. savings and loan associations, credit unions, savings banks; total deposits
10. supplies fiduciary currency, performs check clearing and collection, holds reserves of depository institutions, acts as the government's fiscal agent, supervises depository institutions, acts as lender of last resort, regulates the money supply, intervenes in foreign currency markets; the regulation of the money supply
11. rises; falls
12. saving; investors
13. adverse selection
14. 12

TRUE-FALSE QUESTIONS

1. F That statement is true for a barter economy.
2. T
3. T
4. F The opportunity cost of holding money is foregone interest.
5. F M2 includes non-checkable deposits (near monies).
6. F Barter economies require a double coincidence of wants, and they limit opportunities for specialization.
7. F The U.S. is on a fiduciary monetary standard.
8. T
9. F As discussed in the text, there are a number of barter sites on the Web.
10. F They are more liquid.
11. F They are all highly liquid, but some are not a medium of exchange.
12. T
13. T
14. F Checkable deposits are.
15. F Its most important function is to regulate the money supply.
16. F That is a *function* of the Fed's.
17. T
18. F It is the process of transferring household saving to business investment.
19. F This is a situation of adverse selection.

MULTIPLE CHOICE QUESTIONS

1.d;   2.c;   3.a;   4.d;   5.c;   6.b;   7.a;   8.b;   9.b;   10.d;
11.b; 12.a; 13.c; 14.b; 15.b; 16.b; 17.b; 18.a.

MATCHING

a and j; b and k; c and i; d and h; f and g;   e and l

PROBLEMS

1. a. M1 = $1,272.2 billion
   b. M2 = $6,855.5 billion

GLOSSARY TO CHAPTER 14

Adverse selection The likelihood that individuals who seek to borrow money may use the funds that they receive for unworthy, high-risk projects.

Assets Amounts owned; all items to which a business or household holds legal claim.

Asymmetric information Possession of information by one party in a financial transaction but not by the other party.

Barter The direct exchange of goods and services for other goods and services without the use of money.

Capital controls Legal restrictions on the ability of a nation's residents to hold and trade assets denominated in foreign currencies.

Central bank A banker's bank, usually an official institution that also serves as a country's treasury's bank. Central banks normally regulate commercial banks.

Certificate of deposit A time deposit with a fixed maturity date offered by banks and other financial institutions.

Checkable deposits Any deposits in a thrift institution or a commercial bank on which a check may be written.

Depository institutions Financial institutions that accept deposits from savers and lend those deposits out at interest.

Fiduciary monetary system A system in which currency is issued by the government and its value is based uniquely on the public's faith that the currency represents command over goods and services.

Financial intermediaries Institutions that transfer funds between ultimate lenders (savers) and ultimate borrowers.

Financial intermediation The process by which financial institutions accept savings from businesses, households, and governments and lend the savings to other businesses, households, and governments.

International financial diversification Financing investment projects in more than one country.

Liabilities Amounts owed; the legal claims against a business or household by nonowners.

Liquidity The degree to which an asset can be acquired or disposed of without much danger of any intervening loss in *nominal value* and with small transaction costs. Money is the most liquid asset.

Liquidity approach A method of measuring the money supply by looking at money as a temporary store of value.

M1 The money supply, taken as the total value of currency plus checkable deposits plus traveler's checks not issued by banks.

M2 M1 plus (1) savings and small-denomination time deposits at all depository institutions, (2) balances in retail money market mutual funds, and (3) money market deposit accounts (MMDAs).

Medium of exchange Any asset that sellers will accept as payment.

Money Any medium that is universally accepted in an economy both by sellers of goods and services as payment for those goods and services and by creditors as payment for debts.

Money market deposit accounts Accounts issued by banks yielding a market rate of interest with a minimum balance requirement and a limit on transactions. They have no minimum maturity.

Money market mutual funds Funds of investment companies that obtain money from the public, that are held in common and used to acquire short maturity credit instruments, such as certificates of deposit and securities sold by the U.S. government.

Money supply The amount of money in circulation.

Moral hazard The possibility that a borrower might engage in riskier behavior after a loan has been obtained.

Near monies Assets that are almost money. They have a high degree of liquidity; they can be easily converted into money without loss in value. Time deposits and short-term U.S. government securities are examples.

Savings deposits Interest-earning funds that can be withdrawn at any time without payment of a penalty.

Standard of deferred payment A property of an asset that makes it desirable for use as a means of settling debts maturing in the future; an essential property of money.

Store of value The ability of an item to hold value over time; a necessary property of money.

The Fed The Federal Reserve System; the central bank of the United States.

Thrift institutions Financial institutions that receive most of their funds from the savings of the public; they include savings banks, savings and loan associations, and credit unions.

Time deposit A deposit in a financial institution that requires a notice of intent to withdraw or must be left for an agreed period. Withdrawal of funds prior to the end of the agreed period may result in a penalty.

Transactions accounts Checking account balances in commercial banks and other types of financial institutions, such as credit unions and savings banks; any accounts in financial institutions on which you can easily write checks without many restrictions.

Transactions approach A method of measuring the money supply by looking at money as a medium of exchange.

Traveler's checks Financial instruments purchased from a bank or a nonbanking organization and signed during purchase that can be used as cash upon a second signature by the purchaser.

Unit of accounting A measure by which prices are expressed; the common denomination of the price system; a central property of money.

Universal banking Environment in which banks face few or no restrictions on their powers to offer a full range of financial services and to own shares of stock in corporations.

World index fund A portfolio of bonds issued in various nations whose yields generally move in offsetting directions, thereby reducing the overall risk of losses.

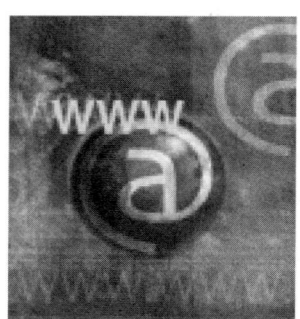

CHAPTER 15
MONEY CREATION AND DEPOSIT INSURANCE

LEARNING OBJECTIVES

After you have studied this chapter, you should be able to

1. define money multiplier, fractional reserve system, total reserves, legal reserves, required reserve ratio, required reserves, excess reserves, balance sheet, open market operations, discount rate, federal funds market, federal funds rate;

2. distinguish among legal, required, and excess reserves;

3. calculate required reserves given the required reserve ratio and total deposits subject to reserve requirements;

4. recognize from a balance sheet whether a bank is in a position to make new loans;

5. distinguish between the situations in which a bank receives a deposit written on another bank and one written on the Fed;

6. show, using balance sheet accounts, what happens when the Fed engages in open market operations;

7. calculate the maximum money multiplier, given the required reserve ratio;

8. recognize forces that reduce the money multiplier;

9. distinguish among the three ways in which the Fed changes the money supply;

10. recognize flaws in the deposit insurance system, and understand how such flaws contribute to depository institution risk-taking; and relate the behavior of the main characters in the S&L crisis to the concepts of adverse selection and moral hazard.

CHAPTER OUTLINE

1. Changes in money supply growth are linked to changes in economic growth, to the inflation rate, and to the business cycle.

2. The Federal Reserve and depository institutions together determine the total money supply in the United States, which has a fractional reserve banking system.

3. Depository institutions are required to maintain a specified percentage of their customer deposits as reserves.
    a. Legal reserves constitute what depository institutions are allowed by law to claim as reserves; today that consists of deposits held at Fed district banks and vault cash.
    b. Required reserves are the minimum amount of legal reserves that a depository institution must hold in reserve; they are expressed as a ratio of required reserves to total deposits.
    c. Excess reserves are the difference between actual (total) legal reserves and required reserves.

4. A balance sheet indicates the relationship between reserves and total deposits in a depository institution.
    a. When a bank receives a deposit drawn on another bank, its total reserves rise by the amount of the deposit, and its excess reserves rise also; the bank lends an amount equal to its excess reserves by creating a checkable account.
    b. The total money supply is unaffected, however, because the bank on which the deposit was drawn will lose an equal number of deposits—and excess reserves—so no net change in total reserves results, and therefore the total money supply is not altered.

5. The Fed can affect the level of reserves in the banking system; therefore, it can change the money supply by a multiple of its transactions.
    a. Open market operations are the buying and selling of U.S. government securities, on the open market, by the Fed.
    b. If the Fed purchases a $100,000 U.S. government security from a bond dealer, it pays for the security by writing a check on itself; when the bond dealer deposits the check, this becomes a bank liability *and* a bank asset/reserve; the bank's excess reserves rise and no other bank's reserves fall; banking system total reserves rise.

6. If the Fed sells a $100,000 U.S. government security to a bond dealer on the open market, the bond dealer pays for it by check, and the Fed reduces the reserves of that bank, while no other bank's reserves have increased; the net effect is a reduction in the banking system's total reserves.

7. When total banking system reserves change, the money supply will change by a multiple of the reserve change.
    a. If the Fed purchases a $100,000 government security, total banking system reserves rise and the money supply will rise by a multiple of $100,000.
    b. Continuing this example, if the required reserve ratio is 20 percent, then total new deposits will equal $500,000.
    c. A financial transaction must increase (decrease) total banking system reserves in order for a multiple expansion (contraction) to occur.

8. The maximum money multiplier equals the reciprocal of the required reserve ratio; the actual change in the money supply equals the actual money multiplier times the change in total reserves.
    a. There are two major forces that reduce the money multiplier.
        i. Currency drains reduce the money multiplier.
        ii. If depository institutions hold excess reserves, the actual money multiplier will be less than the maximum.
    b. The M1 multiplier has varied from about 2.5 to about 3.0 since 1960; the M2 multiplier has increased steadily over that period, from about 6.5 to more than 12.

CHAPTER 15: MONEY CREATION AND DEPOSIT INSURANCE

9. There are three ways in which the Fed can change the money supply: through open market operations, by changing the discount rate, and by changing reserve requirements.
   a. Open market operations are the main way in which the Fed affects the money supply; by buying and selling government securities, the Fed alters the quantity of reserves in the banking system and influences the quantity of deposits in the banking system via the money multiplier.
   b. Although the Fed could in principle conduct monetary policy by changing the differential between the discount rate and the federal funds rate, since 2003 it has maintained the discount rate one percentage point above the federal funds rate.
   c. The Fed rarely changes the required reserve ratio; in addition, sweep accounts, which are accounts with facilities for funds to be shifted from checkable deposits subject to reserve requirements to savings deposits with zero reserve requirements, have greatly reduced banks' effective reserve requirements.

10. Federal deposit insurance was created during the Great Depression in order to prevent bank runs.
    a. There have been three major flaws in the deposit insurance system: the insurance price has been too low; the rate charged has been the same for all institutions, regardless of the riskiness of an institution's portfolio; and the system insures deposits, not depositors.
    b. Those flaws have encouraged managers of depository institutions to assume greater risk than they would otherwise; a moral hazard exists for such managers.
    c. The adverse selection problem also exists; knowing that depositors have little or no incentive to monitor bank manager decisions, people willing to engage in fraudulent and risky behavior are attracted to this industry.
    d. Both the moral hazard problem and the adverse selection problem result from asymmetric information.
    e. Increased risk-taking, in conjunction with adverse economic circumstances, led to large numbers of depository institution failures and governmentally forced mergers.

KEY TERMS
Reserves
Legal reserves
Required reserves
Money multiplier

Required reserve ratio
Excess reserves
Balance sheet
Federal funds market

Open market operations
Discount rate
Net worth
Sweep accounts

KEY CONCEPTS
Money expansion
Money multiplier process
Bank run

Fractional reserve banking system
Federal Deposit Insurance Corporation (FDIC)
Federal funds rate

COMPLETION QUESTIONS
Fill in the blank, or circle the correct term.

1. There is a strong link between growth in the _____ and economic growth, the inflation rate, and the business cycle.

2. Total transactions deposits are determined by the Federal Reserve and _____ institutions.

3. Predecessors to modern-day banks were _____ and money-lenders who had secure vaults and who eventually realized that at any given time only a small _____ of total deposits left with them were withdrawn; this was the beginning of _____ banking.

4. Legal reserves constitute anything that depository institutions are permitted by law to claim as such; today legal reserves include _____ and _____.

5. Required reserves are the (minimum, maximum) amount of legal reserves that a depository institution must hold; the required reserve ratio is the ratio of required reserves to _____; excess reserves equal _____ minus _____.

6. When an individual bank has zero excess reserves, it (can, cannot) extend more loans; when an individual bank receives a new deposit, its total reserves (fall, rise) by the amount of the deposit and its excess reserves (fall, rise). That bank can now increase its lending by the size of _____.

7. If Bank A receives a deposit that is a check written on Bank B, then Bank A's excess reserves will (rise, fall) and Bank B's will _____; overall banking system excess reserves will (rise, fall, remain unaltered); the total money supply will _____.

8. Assume the Fed purchases a $1 million security on the open market from a bond dealer who deposits it in Bank A. Bank A's total reserves rise by $_____ and its excess reserves rise by a fraction of that amount; as a result of this transaction other banks find that their reserves are (decreased, unaltered, increased); the money supply has (decreased, remained constant, increased).

9. If the Fed sells a $1 million security on the open market, total banking system reserves will (fall, rise), and the _____ will contract by a multiple of $1 million.

10. If the required reserve ratio is 10 percent, then the maximum money multiplier equals _____; the actual money multiplier will be less than the maximum due to _____ and _____.

11. If Mr. Calvo deposits a $200 check in Bank A, the $200 increase in Mr. Calvo's checking account is (an asset, a liability) of Bank A; Bank A's (assets, liabilities) will also rise by $200 because Bank A's reserves rise by that amount.

12. The federal deposit insurance system was created during the _____; it was instituted to prevent bank _____.

13. The federal deposit insurance system has three flaws: the rate charged to depository institutions has been too (low, high), each depository institution has been charged (the same, a different) rate, and the system insures (deposits, depositors).

14. Because individual depository institutions don't pay rates that reflect the riskiness of their portfolios, Federal deposit insurance (encourages, discourages) risk-taking by depository institutions.

15. Both the moral hazard problem and the adverse selection problem stem from a(n) _____ of information. The moral hazard problem results because one party to the transaction has superior information (before, after) the transaction, while the adverse selection problem results because one party has superior information (before, after) the transaction.

16. Our deposit insurance system encourages depositors to (ignore, monitor closely) the loan portfolio of depository institutions; hence bank managers have an incentive to take on (less, more) risk, and the _____ problem results.

17. Our deposit insurance system encourages depositors to (ignore, monitor closely) the behavior of bank owners and bank managers. Hence people who are willing to take excessive risk or people who are willing to engage in fraudulent behavior are (discouraged from, encouraged to) enter(ing) the banking industry, and the _____ problem results.

TRUE-FALSE QUESTIONS
Circle the **T** if the statement is true, the **F** if it is false. Explain to yourself why a statement is false.

T F 1. Together, the Fed and depository institutions determine the total money supply.

T F 2. Early goldsmiths discovered that at any given time only a small percentage of people who left gold with them for safekeeping asked for their gold.

T F 3. Legal reserves equal required reserves plus excess reserves.

T F 4. Today legal reserves include only vault cash for banks.

T F 5. Total deposits multiplied by the required reserve ratio equals excess reserves.

T F 6. If an individual bank has excess reserves, it can make loans.

T F 7. When a depository institution makes a loan, it creates checkable deposits.

T F 8. When Mr. Plick deposits a $1,000 check in his checkable account in Bank A, Bank A's assets and liabilities each rise by $1,000.

T F 9. The Fed pays interest to depository institutions that hold reserves with Federal Reserve district banks.

T F 10. Depository institutions have an incentive to minimize their excess reserves.

T F 11. If a financial transaction increases total reserves in the banking system, the money supply will rise.

T F 12. The actual money multiplier will equal the maximum money multiplier, if banks hold excess reserves.

T F 13. When people receive checks, they deposit the whole check; they never withdraw part in currency.

T F 14. The Fed can cause an increase in the money supply by lowering the required reserve ratio.

T F 15. A money multiplier exists due to a fractional reserve system.

T F 16. The maximum money multiplier equals the reciprocal of the marginal propensity to save.

T F 17. A sweep account arrangement allows a commercial bank to shift a depositor's funds from savings deposits to checkable deposits to reduce its required reserves.

T F 18. Federal deposit insurance rates have been zero for most U.S. depository institutions in recent years.

T F 19. If the government subsidizes failing banks, bank managers have less incentive to avoid risk or to be efficient.

T F 20. There is a link between the inflation rate and the rate of growth of the money supply.

T F 21. The adverse selection problem implies that if depositors have little or no incentive to monitor bank behavior, some unscrupulous people will attempt to perform banking services.

T F 22. The adverse selection problem, but not the moral hazard problem, stems from an asymmetric information situation.

T F 23. The moral hazard problem results from asymmetric information that exists before a transaction.

MULTIPLE CHOICE QUESTIONS
Circle the letter that corresponds to the best answer.

1. Under a fractional reserve system, depository institutions
    a. cannot keep 100 percent of their deposits on reserve.
    b. must keep 100 percent of their deposits on reserve.
    c. are required to keep a certain percentage of their total deposits on reserve.
    d. cannot hold excess reserves.

2. Early goldsmiths and money-lenders
    a. providing safekeeping of valuables deposited with them.
    b. were the original bankers.
    c. discovered that they could lend out depositors' gold at interest, because only a fraction of gold deposits were requested at any given time.
    d. All of the above

3. If depository institutions were not required to hold reserves,
    a. they would not hold any reserves.
    b. the Fed would have less control over the money supply.
    c. monetary control would be easier for the Fed.
    d. they would hold 100 percent of their deposits on reserve.

4. Today in the United States, depository institution legal reserves include
    a. vault cash and deposits at Federal Reserve district banks.
    b. U.S. government securities.
    c. gold.
    d. All of the above

5. Which of the following is most **UNLIKE** the others?
    a. flow
    b. stock
    c. wealth
    d. balance sheet

6. *Analogy*: the required reserve ratio is to the money multiplier as the _____ is to the national income multiplier.
    a. balance sheet
    b. excess reserve
    c. bond dealer
    d. marginal propensity to save

7. If the Fed purchases $1 million worth of T-bills on the open market, and the required reserve ratio is 10 percent, then
   a. the money supply will increase by $10 million, at a minimum.
   b. the money supply will decrease by $10 million.
   c. the money supply will increase by $10 million, at a maximum.
   d. the money supply will increase by $5 million, at a maximum.

8. If the Fed raises the required reserve ratio from 10 percent to 20 percent, then the maximum money multiplier
   a. will rise from 5 to 10.
   b. will fall from 10 to 5.
   c. will be unaffected.
   d. cannot be calculated, due to leakages.

9. In the United States, excess reserves
   a. earn interest.
   b. always equal zero.
   c. must be positive before lending can occur.
   d. All of the above

10. Legal reserves minus required reserves equal
    a. the required reserve ratio.
    b. actual reserves.
    c. vault cash plus deposits at Fed District Banks.
    d. excess reserves.

11. Which of the following will increase total reserves in the banking system?
    a. Mr. Patullo deposits a check in Bank A, drawn on Bank B.
    b. The Fed sells a security to Mrs. Damson.
    c. Mr. Farano sells a security to the Fed and deposits the check he receives in Bank C.
    d. Mr. Capano withdraws $100 from his checking account.

12. When Mr. McKay deposits a $1000 check in Bank A,
    a. Bank A's reserves (assets) rise by $1000.
    b. Mr. McKay's deposit is an additional liability for Bank A.
    c. Bank A's excess reserves rise and it can increase its lending.
    d. All of the above

13. If Bank B has negative excess reserves, then it
    a. is meeting its reserve requirements.
    b. will call in loans and not relend as old loans are paid off.
    c. must increase its lending.
    d. must shut down.

14. If the required reserve ratio is 10 percent and the Fed sells a $10,000 security on the open market, the money supply will
    a. rise by $10,000 at a minimum.
    b. rise by $100,000 at a maximum.
    c. fall by $100,000 at a maximum.
    d. fall by $10,000.

15. If the required reserve ratio is 10 percent, then
    a. the actual money multiplier is 10.
    b. the maximum money multiplier is 10.
    c. the national income multiplier is 10.
    d. excess reserves equal 0.

16. The actual change in the money supply equals the actual money multiplier
    a. times the change in reserves.
    b. plus the change in reserves.
    c. times the change in excess reserves.
    d. plus the change in excess reserves.

17. Deposit insurance
    a. is unnecessary in a fractional reserve system.
    b. helps to prevent bank runs.
    c. implies that the government should subsidize failing banks.
    d. decreases depository institution risk-taking under the current system.

18. As it now operates, the federal deposit insurance system
    a. subsidizes depository institutions.
    b. charges the same insurance rate to all banks (currently zero for most).
    c. encourages depository institutions to assume more risk.
    d. All of the above

19. Asymmetry of information that exists between federal deposit insurers and the banks they insure
    a. before the institutions receive insurance could lead to a moral hazard problem.
    b. after the institutions receive insurance could lead to an adverse selection problem.
    c. before the institutions receive insurance could lead to an adverse selection problem.
    d. presents no economic problem.

20. The existence of a moral hazard problem in deposit insurance implies that
    a. asymmetry of information exists.
    b. unscrupulous people have an incentive to own and operate banks.
    c. asymmetry of information existed before a transaction.
    d. All of the above

21. Which of the following is an example of moral hazard in deposit insurance?
    a. A group of individuals wish to start a federally insured savings bank in hopes of stealing funds for their own use.
    b. An investment company facing the prospect of bankruptcy applies for a commercial banking license in hopes of obtaining federal insurance.
    c. The manager of a federally insured credit union decides to use depositors' funds to gamble on high-risk bonds issued by the government of a foreign country.
    d. All of the above

MATCHING
Choose an item in Column (2) that best matches an item in Column (1).

(1)	(2)
a. money multiplier	e. moral hazard
b. asymmetry of information	f. maximum money multiplier of 10
c. required reserve ratio	g. reciprocal of required reserve ratio
d. 10 percent required reserve ratio	h. ratio of required reserves to total deposits

CHAPTER 15: MONEY CREATION AND DEPOSIT INSURANCE

PROBLEMS

1. The following table contains several different required reserve ratios that might be imposed on depository institutions. In column 2, calculate the maximum money multiplier for each of the figures given in column 1. In column 3, calculate the maximum amount by which a single depository institution can increase its loans for each dollar of excess reserves on deposit. In column 4, calculate the amount by which the entire banking system can increase deposits for each dollar of excess reserves in the system.

1	2	3	4
12 1/2%	_____	_____	_____
16 2/3%	_____	_____	_____
20%	_____	_____	_____
30%	_____	_____	_____
33 1/3%	_____	_____	_____

2. Below you are given a series of bank balance sheets. Assume that each case presented is independent. Use the information given to post the changes that would result from the specified action in each case. Show reductions using a minus sign and additions with a plus sign.

	Assets	Liabilities
A. A small business writes a check to pay back a $2,000 loan from the same bank.	Reserves: Loans & Securities:	Checking Deposits:
B. The bank makes a $500 loan to you and credits your checking account.	Reserves: Loans & Securities:	Checking Deposits:
C. The bank sells $500 in Treasury Bills to the Fed to make up a reserve deficiency.	Reserves: Loans & Securities:	Checking Deposits:
D. You cash a $25 check (at your bank) for date money.	Reserves: Loans & Securities:	Checking Deposits:

3. Suppose you have the balance sheets for three banks, as given in the table below. Assume that the reserve requirement is 20 percent.

   a. Compute the required reserves and place these in row A.
   b. Compute the excess reserves and place these in row B.
   c. Compute the amount of new loans each bank can extend in a multibank system and place these in row C.
   d. Compute the amount of new loans each bank could extend if each were a monopoly bank and place these in row D (Note: This is equivalent to viewing each of the balance sheets below as the balance sheet for the entire banking system.)

Assets	1	2	3
Reserves	$ 5,000	$ 6,000	$ 6,000
Loans	10,000	10,000	10,000
Securities	5,000	6,000	7,000
Liabilities			
Checking deposits	17,500	20,000	18,000
Net worth	2,500	2,000	5,000
A. Required Reserves	_____	_____	_____
B. Excess Reserves	_____	_____	_____
C. New Loans (single bank)	_____	_____	_____
D. New Loans (all banks in the system)	_____	_____	_____

4. Assume a 5 percent required reserve ratio, zero excess reserves, no cash drain, and a ready loan demand. The Fed buys a $1 million T-bill from a depository institution.
   a. What is the maximum money multiplier?
   b. By how much will total deposits rise?

5. The Fed purchases a $1 million T-bill from Mr. Mondrone, who deposits it in Bank 1. Using T-accounts, show the immediate effects on this transaction on the Fed and Bank 1.

6. Continuing the example from problem 5:
   a. Indicate Bank 1's position more precisely using a balance sheet account, if required reserves equal 5 percent of checkable deposits.
   b. By how much can Bank 1 increase its lending?

7. Suppose you are the president of Hometown Bank, a small commercial bank. You have recently hired a new loan officer with little experience. Your bank is "loaned up"—in other words, you have no excess reserves. Your new loan officer comes to you very excited and explains that one of your customers has just deposited $5,000 in the bank, so he just approved an auto loan for another customer who was at his desk when the new deposit was made. The loan was for $5,000.

   Can you explain to your new employee why this loan might get the bank into difficulty?

ANSWERS TO CHAPTER 15

COMPLETION QUESTIONS

1. money supply
2. depository
3. goldsmiths; fraction; fractional reserve
4. vault cash; reserves at Federal Reserve district banks
5. minimum; total deposits; actual (total) legal reserves; required reserves
6. cannot; rise; rise; its excess reserves
7. rise; fall; remain unaltered; remain constant
8. $1 million; unaltered; increased
9. fall; money supply
10. 10; cash drains; excess reserves
11. liability; assets
12. 1930s' Great Depression; runs
13. low, the same, deposits
14. encourages
15. asymmetry; after; before
16. ignore; more; moral hazard
17. ignore; encouraged to; adverse selection

TRUE-FALSE QUESTIONS

1. T
2. T
3. T
4. F   Deposits held at Federal Reserve district banks also count as legal reserves.
5. F   That product equals required reserves.
6. T
7. T
8. T
9. F   Reserves do not earn interest.
10. T
11. T
12. F   Excess reserves lower the actual money multiplier.
13. F   They often withdraw cash.
14. T
15. T
16. F   It equals the reciprocal of the required reserve ratio.
17. F   Sweep accounts shift funds from checkable deposits to savings deposits.
18. T
19. T
20. T
21. T
22. F   Both result from asymmetric information.
23. F   For the moral hazard problem, the asymmetric information exists *after* the transaction.

MULTIPLE CHOICE QUESTIONS

1.c;   2.d;   3.b;   4.a;   5.a;   6.d;   7.c;   8.b;   9.c;   10.d;
11.c; 12.d; 13.b; 14.c; 15.b; 16.a; 17.b; 18.d; 19.c; 20.a;
21.c.

MATCHING

a and g;   b and e;   c and h;   d and f

PROBLEMS

1. column 2: 8, 6, 5, 3.33, 3;   column 3: $1, $1, $1, $1, $1
   column 4: $8, $6, $5, $3.33, $3

CHAPTER 15: MONEY CREATION AND DEPOSIT INSURANCE

2.

	Assets	Liabilities
A.	Reserves: 0 Loans & Securities: -$2000	Checking Deposits: -$2000
B.	Reserves: 0 Loans & Securities: +$500	Checking Deposits: +$500
C.	Reserves: +$500 Loans & Securities: -$500	Checking Deposits: 0
D.	Reserves: -$25 Loans & Securities: 0	Checking Deposits: -$25

3. a. Required reserves: $3500, $4000, $3600
   b. Excess reserves: $1500, $2000, $2400
   c. New Loans: $1500, $2000, $2400
   d. New Loans: $7500, $10,000, $12,000

4. a. 20
   b. $20 million

5.

The Fed		Bank 1	
Assets	Liabilities	Assets	Liabilities
+$1,000,000 U.S. govt. securities	+$1,000,000 depository institution reserves	+$1,000,000 reserves	+$1,000,000 checking deposits owned by Mr. Mondrone

6. a.

Bank 1

Assets		Liabilities	
Total reserves Required reserves + (50,000) Excess reserves + ($950,000) Total	+$1,000,000 +$1,000,000	Checking deposits Total	+$1,000,000 +$1,000,000

b. $950,000

7. In order to illustrate the effects of our eager loan officer's actions, we may look at our bank's balance sheet. Initially, we can see no problems, as shown by the balance sheet below, after the new deposit and the loan.

Hometown Bank

Assets		Liabilities	
		Checking Deposits	
Reserves	+ $5,000	Depositors	+ $5,000
Loans	+ 5,000	Borrower	+ 5,000
Total	+ 10,000		+ 10,000

Deposits have been broken down by source so we can see more clearly what happens. As the balance sheet now stands, both reserves and checking deposits have increased by $5,000 as a result of the deposit, and no problems occur.

Presumably, the borrower wishes to purchase a car with the $5,000 loan. As a result, the borrower will probably write a check to the auto dealer. If the auto dealer uses another bank to hold its deposits, then after the auto purchase and the check clearing, our bank's balance sheet would appear as below.

Hometown Bank

Assets		Liabilities	
		Checking Deposits	
Reserves	$ 0	Depositor	+ $5,000
Loans	+ 5,000	Borrower	0
Total	+ 5,000		+ 5,000

Since we were told in the question that the bank was "loaned up," we can see from the balance sheet that our bank now has no reserves with which to cover the latest deposits. Indeed, our eager loan officer has put our bank in a position of a reserve deficiency. Our bank must obtain reserves to meet the legal reserve requirement. This may be done by borrowing reserves, liquidating loans, or by not relending as old loans mature.

GLOSSARY TO CHAPTER 15

Balance sheet A statement of the assets and liabilities of any business entity, including financial institutions and the Federal Reserve System. Assets are what is owned; liabilities are what is owed.

Bank runs Attempts by many of a bank's depositors to convert checkable and time deposits into currency out of fear that the bank's liabilities may exceed its assets.

Discount rate The interest rate that the Federal Reserve charges for reserves that it lends to depository institutions. It is sometimes referred to as the *rediscount rate* or, in Canada and England, as the *bank rate*.

Excess reserves The difference between legal reserves and required reserves.

Federal Deposit Insurance Corporation (FDIC) A government agency that insures the deposits held in banks and most other depository institutions; all U.S. banks are insured this way.

Federal funds market A private market (made up mostly of banks) in which banks can borrow reserves from other banks that want to lend them. Federal funds are usually lent for overnight use.

Federal funds rate The interest rate that depository institutions pay to borrow reserves in the interbank federal funds market.

Fractional reserve banking A system in which depository institutions hold reserves that are less than the amount of total deposits.

Legal reserves Reserves that depository institutions are allowed by law to claim as reserves—for example, deposits held at Federal Reserve district banks and vault cash.

Money multiplier The reciprocal of the required reserve ratio, assuming no leakages into currency and no excess reserves. It is equal to 1 divided by the required reserve ratio.

Net worth The difference between assets and liabilities.

Open market operations The purchase and sale of existing U.S. government securities (such as bonds) in the open private market by the Federal Reserve System.

Required reserve ratio The percentage of total deposits that the Fed requires depository institutions to hold in the form of vault cash or deposits with the Fed.

Required reserves The value of reserves that a depository institution must hold in the form of vault cash or deposits with the Fed.

Reserves In the U.S. Federal Reserve System, deposits held by district Federal Reserve banks for depository institutions, plus depository institutions' vault cash.

Sweep account A depository institution account that entails regular shifts of funds from transaction deposits that are subject to reserve requirements to savings deposits that are exempt from reserve requirements.

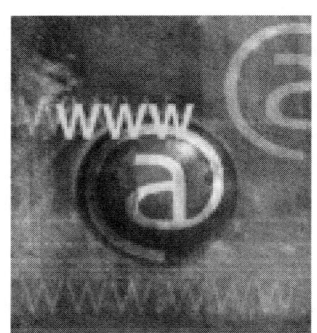

CHAPTER 16
DOMESTIC AND INTERNATIONAL DIMENSIONS OF MONETARY POLICY

LEARNING OBJECTIVES

After you have studied this chapter, you should be able to

1. define transactions demand, precautionary demand, asset demand, monetarism, monetary rule, monetary policy targets, and income velocity of money;

2. list the three reasons for holding money and recognize the main determinant of each;

3. enumerate the three main tools of monetary policy and show an understanding of how they work;

4. determine what happens to the price of a bond when the market interest rate changes;

5. recognize how changes in the money supply affect interest rates, investment spending, equilibrium real national income, and employment in the Keynesian model, and recognize how monetary policy works in the aggregate supply/aggregate demand model and in the equation of exchange approach;

6. use the equation of exchange to determine what happens to the price level when the money supply changes;

7. recognize the main tenets of the monetarist school and evaluate reasons why they favor a monetary rule;

8. list problems involved in the conduct of monetary policy;

9. recognize the main issues involved in setting interest rate versus money supply targets;

10. recognize specific disagreements between the monetarists and the Keynesians;

11. predict the effect on the price level when the supply of money changes relative to the demand for money.

CHAPTER OUTLINE

1. Monetary policy is the Fed's changing of the money supply (or the rate at which it grows) in order to achieve national economic goals.

2. People want to hold money (hence we analyze the demand for money) in order to make foreseen transactions (the transactions demand for money), to make unforeseen expenditures and meet emergencies (precautionary demand for money), and to have a store of value (asset demand for money).
   a. The opportunity cost of holding money is foregone interest earnings.
   b. The demand curve for money is negatively sloped because there is a trade-off between the benefits to holding money and the costs of holding money; in short, as the (opportunity) cost of holding money rises at higher interest rates, people want to hold less money.

3. When the money supply is increased there is a direct and an indirect effect on the economy.
   a. Money supply increases (decreases) directly lead people to spend more (less) because they now have an excess (a shortage) of money balances.
   b. Not all of the excess balances will be spent on goods and services; some excess balances will be used to purchase interest-earning assets, which will cause the interest rate to fall; thus, indirectly, spending will increase on business investment and on consumer durables.
   c. In the real world the money supply usually rises regularly; hence, policy deals with changes in the rate of growth of the money supply.

4. In the long run the higher price level that results from a rightward shift in the AD curve (due to an increase in the money supply) will generate an upward shift in the SRAS curve; ultimately the economy will operate on the LRAS curve, at a higher price level.

5. The Fed has three tools at its disposal when it conducts monetary policy.
   a. Open market operations occur when the Fed buys (sells) bonds in order to increase (decrease) the money supply.
      i. In order to induce people to buy (sell) bonds, the Fed must offer to sell them at a lower (offer to buy them at a higher higher) price; changes in bond prices lead to opposite direction changes in interest rates.
      ii. When the Fed purchases (sells) bonds on the open market, depository institution reserves rise (fall) and such institutions increase (decrease) their lending by creating (destroying) deposits—money.
   b. In principle, the Fed can change the discount rate on loans to depository institutions; a lower (higher) discount rate induces depository institutions to borrow more (less) from the Fed which increases (decreases) the institutions' reserves and leads to more (less) lending and an increase (a decrease) in the money supply.
   c. Although it does so infrequently, when the Fed changes the reserve requirement of depository institutions, it creates or reduces excess reserves, which ultimately changes the money supply, or the rate at which it grows.

6. An expansionary monetary policy, if it reduces interest rates, will cause net exports to rise—an effect which is the opposite of an expansionary fiscal policy (which may cause interest rates to rise).

7. An abundance of empirical evidence, and economic theory, suggest that if the supply of money is consistently increased relative to the demand for money, the relative price of money will fall continuously—inflation will ensue.

8. The equilibrium rate of interest is determined in the market for money.
   a. The total demand for money curve is negatively sloped.

b. The equilibrium interest rate is where the total demand for money curve intersects the money supply curve; at any other interest rate, a surplus or a shortage of money exists.
c. In this model, a change in the money supply (given the demand for money) changes the interest rate, which changes autonomous investment spending. This change in autonomous investment spending shifts the planned expenditures curve, which changes national income by the multiplier effect.
d. Monetarism is the modern quantity theory of money.
   i. The tenets of monetarism are: significant money supply changes lead to significant price level changes in the same direction; money supply changes affect national output and employment only in the short run, but affect only the price level in the long run; fiscal policy is ineffective; the monetary time lags are long and variable, which makes monetary policy difficult to conduct; policymakers should follow a monetary rule instead of using their own discretion.
   ii. In this model, money supply changes upset the community's equilibrium regarding its wealth portfolio; people substitute among bonds, money, equities, and durable goods as the Fed changes the money supply.

9. Some policymakers wish to conduct monetary policy by selecting and meeting interest rate targets, while others wish to attain money stock growth targets.
   a. The Fed cannot, in a meaningful way, target interest rates and the money supply at the same time.
   b. In general, if the demand for money is relatively stable, the Fed should target the money supply; if the demand for money is less stable than private and public expenditures, then the Fed should target interest rates.

KEY TERMS
Asset demand
Transactions demand

Precautionary demand
Demand for money

KEY CONCEPTS
Monetary rule
Crude quantity theory of money and prices
Monetarists

Equation of exchange
Income velocity of money

COMPLETION QUESTIONS
Fill in the blank, or circle the correct term.

1. The Fed (can, cannot) control the supply of money, but it (can, cannot) control the demand for money.

2. The _____ theory of money and prices predicts that changes in the price level are determined by changes in the quantity of money in circulation.

3. If the supply of money rises relative to its demand, the price level will (rise, fall) and the value of a unit of money will (rise, fall).

4. The number of times, on average, that each monetary unit is spent on final goods and services is called the _____.

5. In the crude quantity theory of money model, in the long run national output (will, will not) be at full employment, and velocity (is, is not) constant; hence if the money supply doubles, the price level will _____.

6. People hold money for three motives: _____, _____, and _____.

7. The transactions demand for money varies (directly, inversely) with nominal national income; the asset demand for money varies (directly, inversely) with the interest rate.

8. When the interest rate rises, bond prices (rise, fall); when bond prices rise, the interest rate _____; the opportunity cost of holding money is foregone _____.

9. From a Keynesian perspective, a fall in the interest rate causes autonomous net investment to (rise, fall), which in turn causes the aggregate demand curve to shift (rightward, leftward); then _____ and employment will rise.

10. If the current interest rate is below the equilibrium rate, an excess _____ exists and the community will attempt to (buy, sell) bonds, thereby forcing the price of bonds _____ and the interest rate _____.

11. Monetarists maintain that full employment is normal in the long run. They believe the following: that the income velocity of money (is, is not) stable; that changes in the money supply lead to changes in the price level in the (same, opposite) direction; that money supply changes affect _____ and _____ in the short run; but that in the long run only the _____ is affected by money supply changes.

12. Monetarist critics of the Fed (do, do not) want the Fed to pursue discretionary monetary policy; they want the Fed to follow a(n) _____.

13. The three main tools the Fed employs to conduct monetary policy are _____, _____, and _____.

14. When the Fed buys bonds on the open market, the price of bonds tends to (fall, rise) and the interest rate (falls, rises); also bank reserves (fall, rise) which leads to (a decrease, an increase) in bank lending; ultimately the money supply (falls, rises).

15. If the Fed raises the differential between the discount rate and the federal funds rate, this will (discourage, encourage) banks from borrowing from it and may (decrease, increase) bank reserves, leading to (a decrease, an increase) in bank deposit creation; the money supply will (fall, rise) and interest rates will probably (fall, rise) in the short run.

16. If the Fed raises reserve requirements, banks will find it (harder, easier) to meet reserve requirements; banks will (call in loans, lend more) and the money supply will (fall, rise).

17. An expansionary monetary policy will tend to cause the interest rate to fall, which may lead to (a decrease, an increase) in net exports; an expansionary fiscal policy financed by deficits may cause the interest rate to (fall, rise), which may lead to (a decrease, an increase) in net exports.

TRUE-FALSE QUESTIONS
Circle the **T** if the statement is true, the **F** if it is false. Explain to yourself why a statement is false.

T F 1. An important objective of monetary policy is to assist the economy in maintaining high employment without undue inflation.

T F 2. The Fed can control the supply of money but not the demand for money.

T F 3. The crude quantity theory of money maintains that if the money supply is doubled, the price level is halved.

T F 4. The asset demand for money motive stresses money's role as a medium of exchange.

T F 5. The asset demand for money motive stresses money's role as a liquid store of value.

T F 6. The opportunity cost of holding money is loss of liquidity.

T F 7. When interest rates rise, the price of existing bonds falls.

T F 8. According to the Keynesian view, the supply of and the demand for money directly determine net investment spending.

T F 9. If an excess quantity demanded for money exists, people will attempt to sell bonds, which drives interest rates up.

T F 10. If the Fed buys bonds on the open market, bank reserves will rise, and so will bank lending.

T F 11. The Fed cannot target both the interest rate and the money supply.

T F 12. If the demand for money is more stable than private expenditures, the Fed should target the money supply.

T F 13. If the Fed wants to increase the money supply, in principle it can raise the discount rate or raise reserve requirements.

T F 14. Monetarists believe that discretionary monetary policy is effective, but that fiscal policy is not.

T F 15. An expansionary monetary policy tends to increase a nation's net exports, while an expansionary fiscal policy tends to decrease its net exports.

T F 16. Keynesian economists prefer a monetary rule to discretionary fiscal policy because they believe that the demand for money is unstable.

MULTIPLE CHOICE QUESTIONS
Circle the letter that corresponds to the best answer.

1. The demand for money curve
   a. is upward sloping.
   b. is flat.
   c. is downward sloping.
   d. denies a link between the money supply and the price level.

2. In the equation $M_S V = PY$, according to the crude quantity theory,
   a. M_S is independent of the price level.
   b. V is the number of times each dollar is spent, on average, per year.
   c. Y is the real price level.
   d. P rises as V falls, other things constant.

3. In the crude quantity theory, as the money supply rises (other things constant),
   a. the quantity demanded for money rises.
   b. velocity rises.
   c. velocity falls.
   d. the price level rises proportionately.

CHAPTER 16: DOMESTIC AND INTERNATIONAL DIMENSIONS OF MONETARY POLICY

4. Which of the following stresses money's role as a medium of exchange?
   a. transactions demand
   b. precautionary demand
   c. asset demand
   d. miserly demand

5. Which of the following stresses money's role as a liquid store of value?
   a. transactions demand
   b. precautionary demand
   c. asset demand
   d. miserly demand

6. The opportunity cost of holding money is
   a. foregone liquidity.
   b. foregone interest income.
   c. convenience.
   d. security.

7. The demand for money curve
   a. reflects a preference for money over bonds.
   b. shows a negative relationship between the quantity demanded for money and the interest rate.
   c. shows that people want to substitute money for bonds at low interest rates.
   d. All of the above

8. When interest rates rise,
   a. bond prices rise.
   b. bond holders experience capital losses.
   c. bond prices are unaffected.
   d. bond holders experience capital gains.

9. An excess quantity supplied of money
   a. exists at all interest rates above equilibrium.
   b. causes interest rates to rise.
   c. induces people to want to sell bonds.
   d. All of the above

10. The intersection of the supply of and the demand for money determines the
    a. aggregate demand curve.
    b. planned expenditures curve.
    c. price level.
    d. interest rate.

11. An expansionary monetary policy is beneficial if
    a. the unemployment rate is relatively high.
    b. a recession exists.
    c. the price level is falling.
    d. All of the above

12. According to the Keynesian perspective, an increase in the money supply will
    a. reduce the interest rate.
    b. reduce net investment spending.
    c. reduce total planned expenditures.
    d. All of the above

13. Which of the following clearly is **NOT** a result of an expansionary monetary policy?
    a. higher price level
    b. lower inflation rate
    c. increase in nominal national income
    d. increased total expenditures

14. A contractionary monetary policy
    a. shifts the AD curve rightward.
    b. shifts the AD curve leftward.
    c. shifts the LRAS curve, but not the AD curve.
    d. None of the above

15. If the velocity of money is less stable than private expenditures, then the Fed
    a. should target the interest rate.
    b. should pursue fiscal policy.
    c. should target a monetary aggregate.
    d. cannot target the interest rate.

16. The equation of exchange states that
    a. expenditures equal receipts.
    b. spending equals saving.
    c. saving equals investment.
    d. aggregate demand exceeds aggregate supply.

17. According to the crude quantity theory (assuming V and Y are constant), if the money supply is tripled, the price level will
    a. remain unchanged.
    b. fall.
    c. triple.
    d. more than triple.

18. Traditional Keynesians maintain that
    a. investment is a function of the interest rate.
    b. monetary policy works through changes in the interest rate.
    c. the Fed should target interest rates, not the money supply.
    d. All of the above

19. Which of the following is **NOT** a tenet of monetarism?
    a. Monetary policy is destabilizing.
    b. The Fed should follow a monetary rule, and not use its discretion.
    c. Changes in the money supply affect only the price level, in the long run.
    d. Keynesian multipliers are sufficiently reliable to conduct stabilization policies.

20. In regard to the effectiveness of monetary policy, which of the following is true of the net export effect?
   a. The net export effect tends to reinforce the effects of an expansionary monetary policy action, because the rise in interest rates caused by an expansion in the money supply induces inflows of funds from abroad and reduces the value of the dollar. As a result, foreign spending on U.S. export products increases, which tends to boost real national income somewhat.
   b. The net export effect tends to reinforce the effects of an expansionary monetary policy action, because the fall in interest rates caused by an expansion in the money supply induces outflows of funds from the United States and reduces the value of the dollar. As a result, foreign spending on U.S. exports increases, which tends to raise real national income somewhat.
   c. The net export effect tends to offset the effects of an expansionary monetary policy action, because the rise in interest rates caused by an expansion in the money supply induces inflows of funds from abroad and raises the value of the dollar. As a result, foreign spending on U.S. export products declines, which tends to push down real national income somewhat.
   d. The net export effect tends to offset the effects of an expansionary monetary policy action, because the fall in interest rates caused by an expansion in the money supply induces outflows of funds from abroad and raises the value of the dollar. As a result, foreign spending on U.S. export products declines, which tends to push down real national income somewhat.

MATCHING

Choose the item in Column (2) that best matches an item in Column (1).

(1)	(2)
a. equation of exchange	h. nondiscretionary monetary policy
b. monetarism	i. preference for liquidity
c. velocity	j. monetary aggregate
d. transactions motive	k. crude quantity theory of money
e. monetary target	l. average turnover rate of money supply
f. monetary rule	m. inflation as a monetary phenomenon
g. asset demand	n. money as a medium of exchange

WORKING WITH GRAPHS

1. On the next page in panel (a) is the investment demand function for the economy. Panel (b) represents the supply and demand functions for money. The interest rate in the economy is currently 6 percent. The money supply is $1 trillion. Panel (c) is the Keynesian model of the economy, which is currently at an equilibrium level of output and income of $10 trillion. Full employment output and income is $12 trillion. Further assume that the reserve requirement is 20 percent and there are no excess reserves in the banking system. Also assume that the economy's MPC is 0.6. Autonomous investment is currently $1.8 trillion. Fill in the paragraph that follows the graphs.

   If the goal of the monetary authorities is to reach full employment, the Fed would want to (increase, decrease) the money supply by $_____, which would (raise, lower) interest rates to _____ percent. This would (increase, decrease) the level of autonomous investment by $_____ to $_____. This (increase, decrease) in investment would be subject to a multiplier effect of _____ and therefore increase equilibrium level of income and output by $_____ to $_____. Assuming the Fed chose buying securities on the open market to (increase, decrease) the money supply, how many dollars of securities must the Fed purchase from the nonbank public to (increase, decrease) the money supply that would be consistent with full employment equilibrium? $_____.

184 CHAPTER 16: DOMESTIC AND INTERNATIONAL DIMENSIONS OF MONETARY POLICY

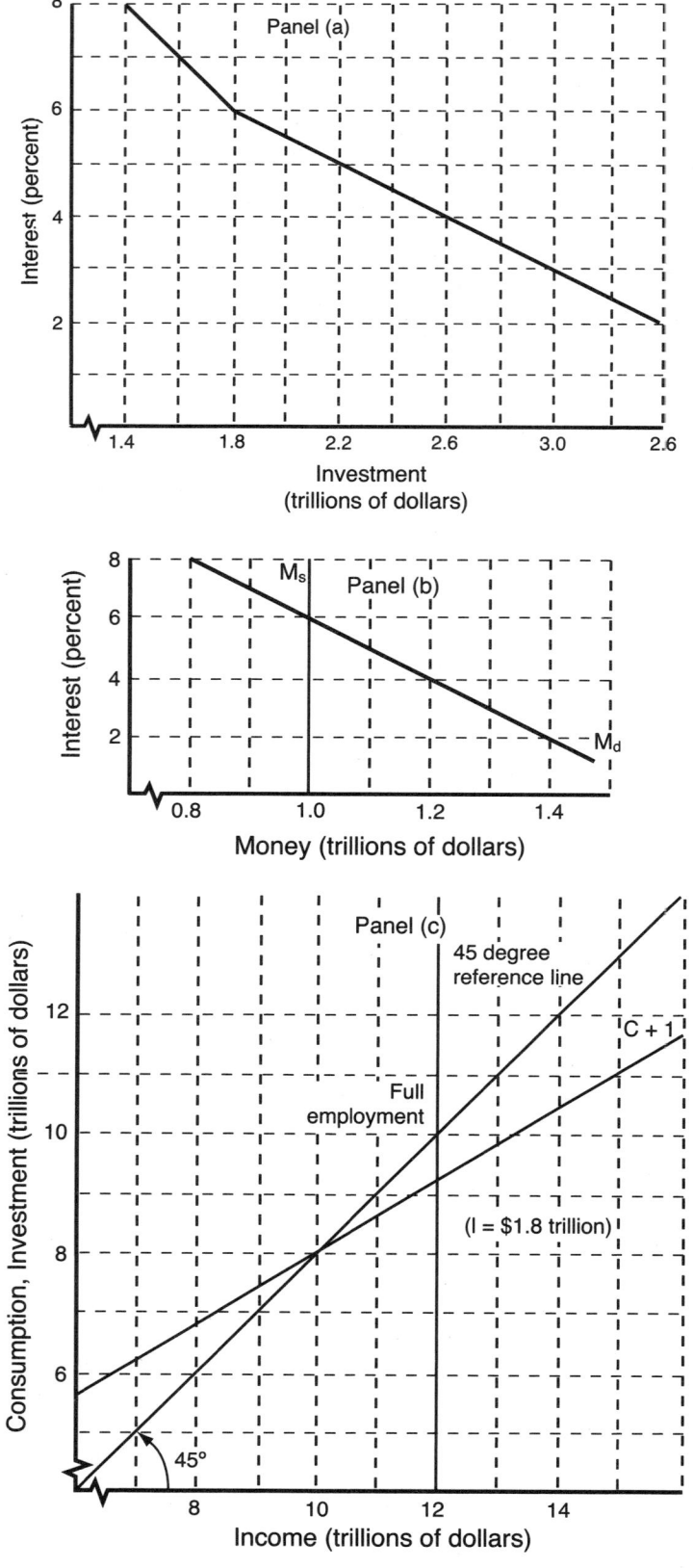

2. Analyze the graphs below, then answer the questions that follow.

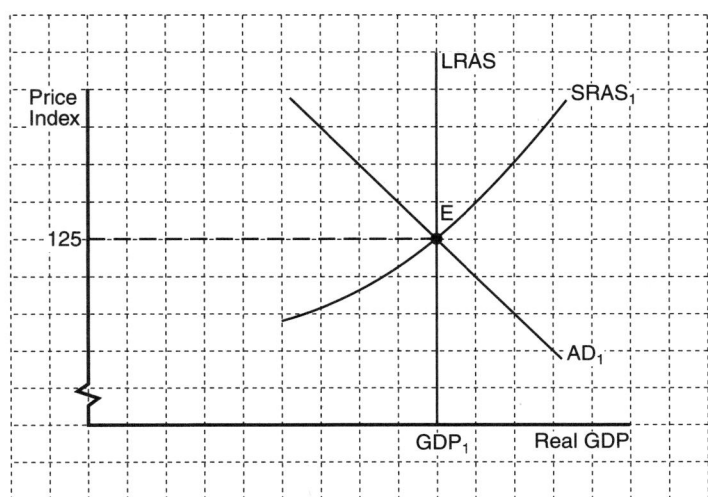

Assume that the economy is operating at point E, and that the Fed wants to reduce the price level.

a. Should it reduce or increase the money supply?
b. Should it buy or sell bonds on the open market?
c. Should it raise or lower the discount rate relative to the federal funds rate?
d. If the Fed follows your advice, what will happen to the AD curve? (Draw it on your book or on a piece of paper.)
e. What will happen to real GDP and the price level? Label the new short-run equilibrium point as A.
f. After resource supplies adjust to the new price level, what happens to the SRAS curve? (Draw it.)
g. Indicate the new position in which both long-run and short-run equilibrium exist by labeling it B.
h. Compare the price level and real GDP level at points E and B.

3. Analyze the graphs below and answer the questions that follow.

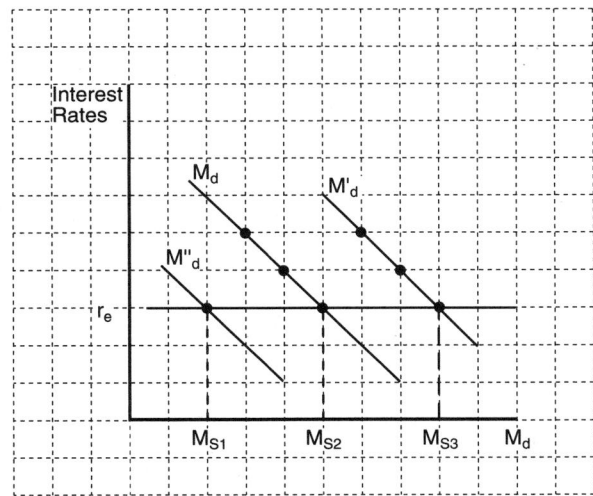

a. If the Fed targets the interest rate at r_e, and if the demand for money shifts leftward (falls) from M_d to M''_d, what happens to the money supply? Is this a stabilizing change in the money supply? Why?

b. If the Fed targets the interest rate at r_e, and if the demand for money shifts rightward (increases) from M_d to M'_d, what happens to the money supply? Is this a stabilizing money supply change? Why?

PROBLEMS

1. Suppose the economic advisors to the president feel it is necessary to raise our national income. These advisors and the president are traditional Keynesians. The chairman of the Fed is called to the Oval Office and told to stimulate the economy. Although the chairman does not have to obey, he decides to accept the directive.

   Below you will find a series of interrelated questions concerning the execution of this directive. Be careful! An early mistake can affect your remaining answers. Circle the correct answer for each question.

   i. The Fed will do which of the following?
      a. sell securities
      b. raise the discount rate
      c. buy securities
      d. raise reserve requirements

   ii. This action will
      a. lower the reserve requirement.
      b. raise the excess reserves of member banks.
      c. cause banks to increase their participation in the IMF.
      d. lower the amount of reserves required that are on deposit with the Fed.

   iii. We know banks are profit maximizers. They will therefore
      a. lend out excess reserves, which create more demand deposits.
      b. lend out excess reserves, which decrease demand deposits.
      c. call in outstanding loans.
      d. a and b

   iv. After the results of question iii,
      a. the money supply will decrease.
      b. the money supply will increase.
      c. the money supply will increase initially but then will return to its previous level.
      d. there will be no effect on the money supply.

   v. If you have answered question iv correctly, what will happen next?
      a. The interest rate will decrease, and investment will decrease.
      b. The interest rate will go up, with no effect on investment.
      c. The interest rate will remain unaffected.
      d. The interest rate will decrease, and investment will increase.

   vi. With the correct answer to question v, we know that
      a. national income will increase by the amount of investment.
      b. national income will decrease by the amount of investment.
      c. national income will change by more than the amount by which investment will change.
      d. national income will change by less than the amount by which investment will change.

vii. The change in national income will be
   a. change in I times $1/(1-MPS)$.
   b. change in I times $1/(1-MPC)$.
   c. change in I times $1/(1-MPC-MPS)$.
   d. change in I times change in the interest rate.

2. Given the equation of exchange $M_sV = PY$, and $M_s = \$50$, $V = 4$, and $Y = \$100$, then
   a. What does P equal?
   b. If M_s rises from $50 to $100, what happens to P?
   c. If M_s falls from $50 to $25, what happens to P?

3. Given the equation of exchange $M_sV = PY$, and $M_s = \$100$, $V = 5$, and PY = nominal national income, then
   a. What is the value of nominal national income?
   b. If price level equals 1, what is the value of real national income?
   c. If the money supply triples, other things constant, what is the value of nominal national income? of real national income?

ANSWERS TO CHAPTER 16

COMPLETION QUESTIONS

1. can; cannot
2. crude quantity
3. rise; fall
4. income velocity of money
5. will; is; double
6. transactions; precautionary; asset
7. directly; inversely
8. fall; falls; interest earnings
9. rise; rightward; real national income
10. quantity demanded for money; sell; downward; upward
11. is; same; national output; employment; price level
12. do not; monetary rule
13. open market operations; changing the discount rate; changing reserve requirements
14. rise; falls; rise; increase; rises
15. discourage; decrease; decrease; fall; rise
16. harder; call in loans; fall
17. increase; rise; decrease

TRUE-FALSE QUESTIONS

1. T
2. T
3. F   The price level will also double.
4. F   It stresses money's role as a store of value.
5. T
6. F   The opportunity cost is foregone interest earnings.
7. T
8. F   They determine the interest rate; hence investment is indirectly determined.
9. T
10. T
11. T
12. T
13. F   No, the Fed would lower them.
14. F   They believe that neither is effective.
15. T
16. F   Monetarists prefer a monetary rule; Keynesians prefer an activist monetary policy.

188 CHAPTER 16: DOMESTIC AND INTERNATIONAL DIMENSIONS OF MONETARY POLICY

MULTIPLE CHOICE QUESTIONS

1.c; 2.b; 3.d; 4.a; 5.c; 6.b; 7.d; 8.b; 9.a; 10.d;
11.d; 12.a; 13.b; 14.b; 15.a; 16.a; 17.c; 18.d; 19.d; 20.b.

MATCHING
a and k; b and m; c and l; d and n; e and j; f and h; g and i

WORKING WITH GRAPHS

1. increase, $0.2 trillion (or $200 billion); lower; 4; increase; $0.8 trillion; $2.6 trillion; increase; 2.5; $2 trillion; $12 trillion; increase; increase; $0.04 trillion (or $40 billion) in securities from the nonbank public. **Note**: This problem actually requires you to work backward in order to obtain the necessary increase in the money supply. That is, you must first find how much of an increase in autonomous investment is required to increase the equilibrium level of income by $2 trillion (change in I x multiplier = change in income). Substitute 2.5 for the multiplier and $2 trillion for the change in income. Now go to panel (a) and find how much the interest rate must fall to increase investment by $0.8 trillion. Then find how much the money supply must increase to lower the interest rate to 4 percent. The answer is $0.2 trillion. If the Fed purchased $0.04 trillion (or $40 billion) of securities from, for example, businesses and security dealers, they would deposit this in their checking accounts, which would increase demand deposits in the banking system by $0.04 trillion. Finally, the money multiplier of 5 would increase the money supply by the necessary $0.2 trillion ($200 billion).

2. a. reduce; b. sell; c. raise; d. It will shift leftward, to AD_2 on the graph below; e. both fall; f. It shifts downward, to $SRAS_2$; g. See the graph below; h. Real GDP is the same; the price level is lower.

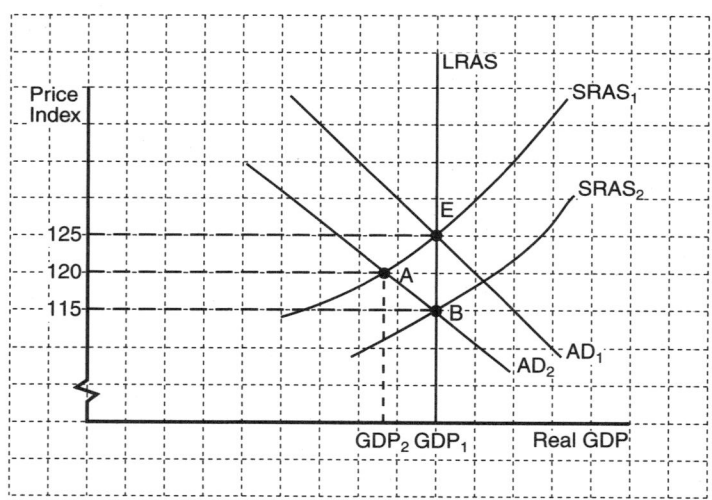

3. a. Decreases to M_{s1}. Yes, because if the demand for money falls and the supply of money falls, then the interest rate will be unaffected and private spending won't change due to a change in the money supply.
   b. Increases to M_s3. Yes, because if the demand for money rises and the supply of money rises, then the interest rate will be unaffected; hence private spending won't change due to an unstable demand for money.

PROBLEMS

1. i. c ii. b iii. a iv. b v. d vi. c vii. b
2. a. P = 2   b. It doubles; P = 4.   c. It is halved; P = 1.
3. a. $500   b. $500   c. $1500; $500

GLOSSARY TO CHAPTER 16

Asset demand Holding money as a store of value instead of other assets such as certificates of deposit, corporate bonds, and stocks.

Crude quantity theory of money and prices The belief that changes in the money supply lead to proportional changes in the price level.

Equation of exchange The formula indicating that the number of monetary units times the number of times each unit is spent on final goods and services is identical to the price level times output (or national income).

Income velocity of money The number of times per year a dollar is spent on final goods and services. It is equal to nominal GDP divided by the money supply.

Monetarists Macroeconomists who believe that inflation in the long run is always caused by excessive monetary growth and that changes in the money supply affect aggregate demand both directly and indirectly.

Monetary rule A monetary policy that incorporates a rule specifying the annual rate of growth of some monetary aggregate.

Precautionary demand Holding money to meet unplanned expenditures and emergencies.

Transactions demand Holding money as a medium of exchange to make payments. The level varies directly with nominal national income.

CHAPTER 17
STABILIZATION IN AN INTEGRATED WORLD ECONOMY

LEARNING OBJECTIVES

After you have studied this chapter, you should be able to

1. define active policymaking, passive policymaking, Phillips curve, nonaccelerating inflation rate of unemployment (NAIRU), rational expectations hypothesis, policy irrelevance proposition, real business cycle approach, small-menu-cost approach, and efficiency wage theory;

2. interpret a Phillips curve;

3. answer questions that indicate an understanding of why the Phillips curve might be negatively sloped in the short run and vertical in the long run;

4. distinguish between the effects of an announced (anticipated) and an unannounced (unanticipated) monetary policy, according to the rational expectations hypothesis;

5. recognize the implications of price and wage inflexibility for small-menu cost theory and efficiency wage theory;

6. recognize the essential features of the real business cycle approach;

7. recognize similarities and distinctions among the alternative arguments supporting active versus passive policy approaches.

CHAPTER OUTLINE

1. Some economists have argued that a trade-off exists between the rate of unemployment and the rate of inflation.
   a. If such a relationship (called a Phillips curve) existed, policymakers could simply try to balance the problems associated with inflation with those associated with unemployment; they could then select the "optimal" mix of unemployment and inflation consistent with societal values.
   b. In reality, no such policy option exists; the Phillips curve relationship has proved to be unstable.

2. A consistent short-run trade-off between inflation and unemployment seems to have existed in the past, when the actual inflation rate was relatively low for long periods.
   a. In recent years the inflation rate has been much more variable.

b. If workers underestimate the true inflation rate, they will be fooled into accepting a lower wage rate than they bargained for, and the actual unemployment rate will fall.

3. The natural rate of unemployment is determined by (a) frictional unemployment and (b) unemployment due to price and wage rigidities in the economic system.

4. The nonaccelerating inflation rate of unemployment (NAIRU) is the rate of unemployment below which the rate of inflation tends to rise and above which it tends to fall.

5. An unanticipated expansionary monetary policy causes the aggregate demand curve to shift rightward while leaving the position of the short-run aggregate supply curve unchanged; hence real national income (output) and employment will rise with the price level.

6. In the long run, expectational adjustments cause the position of the short-run aggregate supply curve to adjust until it crosses the aggregate demand curve along the vertical long-run aggregate supply curve; hence expansionary monetary policy will increase only the price level.

7. The rational expectations hypothesis (REH) contends that economic agents will try to anticipate future stabilization policy and its effects.
    a. Economic agents will allocate resources to predict stabilization policy and its effect on the value of future economic variables—especially the future inflation rate.
    b. The REH predicts that policymakers cannot cause economic agents to make consistent forecasting errors: Workers will not consistently and systematically underestimate the inflation rate; sometimes they will overestimate the inflation rate.
    c. Eventually, once economic agents know that policymakers are trying to fool them, not even a *short-run* (systematic) trade-off between inflation and unemployment exists.
        i. The REH predicts that an anticipated (announced) stabilization policy will have little impact on the employment rate and on real national income.
        ii. Expansionary demand policies in the past may have worked because economic agents had not been fooled by *previous* government stabilization policies.
        iii. In recent years stabilization policy has been more difficult to conduct because economic agents have found it rational to allocate resources to economic forecasting.

8. The policy irrelevance proposition is that policy actions have no real effects in the short run if the policy actions were anticipated, and none in the long run even if the policy actions were unanticipated.
    a. The policy irrelevance proposition follows from combining the rational expectations hypothesis with the assumptions that (a) all markets are highly competitive and (b) all prices and wages are perfectly competitive.
    b. An implication of the policy irrelevance proposition is that fluctuations in real variables are a result of *mistakes* on the part of either policymakers or economic agents.

9. The real business cycle approach maintains the assumption of price and wage flexibility and suggests that the source of business cycles is changes in supply.
    a. Supply shocks—such as the oil supply disruption or the sudden increase in the price of oil in the 1970s, or any significant change in the price of a crucial resource—shift the SRAS curve (and maybe the LRAS curve) and change real economic variables.
    b. Technological changes and changes in the composition of the labor force are additional (nonmonetary) supply shocks that have short-run and long-run effects on real economic variables.

10. Various hypotheses have been developed to explain fluctuations in real variables (business cycles), yet which also support active policymaking.

a. The small-menu cost approach hypothesizes that it is costly for firms to change prices in response to demand changes; hence some price rigidity is rational.
b. The efficiency wage theory maintains that high wages tend to increase labor productivity and worker loyalty, and wage reductions might interfere with both; employers are therefore reluctant to lower wages in recessions, so wage rates are inflexible downward.

KEY TERMS
Active policymaking
Efficiency wage
Nonaccelerating inflation rate of unemployment
Phillips curve

Passive policymaking
Policy irrelevance proposition
Small menu costs

KEY CONCEPTS
Natural rate of unemployment
New Keynesian theory

Rational expectations hypothesis
Real business cycles

COMPLETION QUESTIONS
Fill in the blank, or circle the correct term.

1. The Phillips curve posits a trade-off between the _____ rate and the _____ rate.

2. In recent years in the United States, the predicted Phillips curve relationship (is, is not) supported by empirical evidence.

3. The natural rate of unemployment prevails in the (short, long) run when the economy is in (equilibrium, disequilibrium); when the natural rate of unemployment is reached, the actual inflation rate (is less than, is greater than, equals) the expected inflation rate, and there (is, is not) a tendency for the inflation rate to accelerate.

4. In order to keep the actual unemployment rate below the natural unemployment rate, the actual inflation rate must be (greater than, less than, equal to) the expected inflation rate; thus the inflation rate must always be (constant, accelerating, decelerating).

5. The rational expectations hypothesis contends that policymakers (can, cannot) induce economic agents to make systematic forecasting errors. According to this model, if a government stabilization policy is announced, it will have (much, little) effect on output and unemployment; if a stabilization policy is not announced and is unanticipated by economic agents, it (will, will not) have a short-run impact on the economy, and the impact (will, will not) be systematic on output and employment.

6. If inflationary expectations are high and the Fed pursues an unanticipated contractionary policy, the unemployment rate will (fall, rise) because economic agents will (underestimate, overestimate) the future inflation rate. If the Fed announces a contractionary policy and such a policy is believed by economic agents, the rational expectations hypothesis predicts that the unemployment rate (will, will not) rise significantly.

7. The policy irrelevance proposition follows from adopting the assumption of the rational expectations hypothesis and adding the assumptions of _____ and _____; this proposition implies that fluctuations in real variables are a result of mistakes on the part of _____ and _____.

8. The policy irrelevance proposition states that if policy actions are anticipated, such actions will have (no, a great) effect on real variables in the short run; if policy actions are unanticipated, then they (will, will not) have an effect on real variables in the short run, but

that effect (is, is not) predictable; in the long run, unanticipated policy (will, will not) have real effects.

9. The small-menu-cost approach maintains that if the cost of frequent changes in prices exceeds the costs of not changing such prices, it is (rational, irrational) to leave prices unchanged in the face of changes in demand; this theory suggests that prices and wages (are, are not) perfectly flexible.

10. The efficiency wage theory maintains that (high wages lead to high productivity, high productivity leads to high wages), and therefore producers may be reluctant to reduce wages in recessions; hence this theory suggests that wage rates are (inflexible, flexible) in the downward direction.

11. The real business cycle approach assumes that prices and wages are (inflexible, flexible) and that changes in real economic variables result from (supply, demand) shocks; this theory suggests that changes in _____, _____, and _____ could affect real economic variables.

TRUE-FALSE QUESTIONS
Circle the **T** if the statement is true and the **F** if it is false. Explain to yourself why a statement is false.

T F 1. The Phillips curve relates inflation rates to growth rates.

T F 2. In the short run, equilibrium real GDP per year can be higher than the annual GDP level consistent with the natural rate of unemployment.

T F 3. At the nonaccelerating inflation rate of unemployment, the inflation rate is steady.

T F 4. The nonaccelerating inflation rate of inflation and the natural rate of unemployment are always the same.

T F 5. If the inflation rate is rising at the current rate of unemployment, then the rate of unemployment presently is below the nonaccelerating inflation rate of unemployment.

T F 6. The net export effect arises because an interest-rate reduction generated by an expansionary monetary policy causes a capital outflow that causes the value of the domestic currency to fall, thereby raising aggregate demand and increasing the price level in the short run.

T F 7. Monetary policy works by fooling people only in the long run.

T F 8. The policy irrelevance proposition indicates that if the government announces its stabilization policy, the unemployment rate will be affected greatly.

T F 9. A key assumption of the policy irrelevance proposition is speedy adjustment of wages and prices.

T F 10. In a long-run macroeconomic equilibrium, the actual unemployment rate exceeds the natural unemployment rate.

T F 11. In recent years in the United States, there seems to be no systematic, negative relationship between the inflation rate and the unemployment rate.

T F 12. Under the rational expectations hypothesis, the actual inflation rate always equals the anticipated inflation rate.

T F 13. The rational expectations hypothesis argues that policymakers cannot systematically change the unemployment rate in the short run.

T F 14. According to the policy irrelevance proposition, policymakers simply cannot affect the unemployment rate in the short run.

T F 15. Economists who promote the idea of real business cycles have provided evidence supporting the use of activist policymaking.

T F 16. If the efficiency wage theory is correct, then unemployment for most new workers will be temporary, so the unemployment rate should be relatively low.

T F 17. Small menu costs refer to the costs of renegotiating contracts as well as to the costs of changing price lists, such as restaurant menus.

MULTIPLE CHOICE QUESTIONS
Circle the letter that corresponds to the best answer.

1. Which of the following is most **UNLIKE** the others?
    a. theories that assume fixed prices in the short run
    b. the efficiency wage theory
    c. the real business cycle approach
    d. the small-menu-cost approach

2. The Phillips curve relates
    a. inflation rates and productivity rates.
    b. inflation rates and unemployment rates.
    c. unemployment rates and growth rates.
    d. the natural unemployment rate and the actual unemployment rate.

3. A trade-off between inflation and unemployment
    a. exists only in the long run.
    b. is depicted by the Phillips curve.
    c. is depicted by the production possibilities curve.
    d. is depicted by the long-run aggregate supply curve.

4. The nonaccelerating inflation rate of unemployment is the unemployment rate
    a. at which there is no tendency for inflation to rise or to fall.
    b. that always matches the natural rate of unemployment.
    c. at which the inflation rate rises at a steady pace.
    d. at which the inflation rate falls at a steady pace.

5. Other things being equal, a decrease in the quantity of money in circulation causes a
    a. decline in aggregate demand in the short run and a rightward shift in the short-run aggregate supply curve to bring about a long-run adjustment.
    b. decline in aggregate demand in the short run and a leftward shift in the short-run aggregate supply curve to bring about a long-run adjustment.
    c. leftward shift in the short-run aggregate supply curve in the short run and a leftward movement along the aggregate demand curve to bring about a long-run adjustment.
    d. rightwardward shift in the short-run aggregate supply curve in the short run and a rightward movement along the aggregate demand curve to bring about a long-run adjustment.

6. When domestic interest rates decline following an expansionary monetary policy action,
   a. the fall in the value of the domestic currency that follows an international capital outflow encourages foreign residents to purchase more domestic goods and services, thereby helping to boost aggregate demand and push up the price level in the short run.
   b. the rise in the value of the domestic currency that follows an international capital inflow encourages foreign residents to purchase more domestic goods and services, thereby helping to boost aggregate demand and push up the price level in the short run.
   c. the fall in the value of the domestic currency that follows an international capital outflow discourages foreign residents from purchasing domestic goods and services, thereby helping to depress aggregate demand and push down the price level in the short run.
   d. the rise in the value of the domestic currency that follows an international capital inflow discourages foreign residents from purchasing domestic goods and services, thereby helping to depress aggregate demand and push down the price level in the short run.

7. With regard to a contractionary monetary policy action, the net export effect tends to
   a. work in opposition to the main effect of the policy action by raising the value of the home currency and causing export spending by foreign residents to rise. This helps to raise aggregate demand on net and raises the price level.
   b. work in opposition to the main effect of the policy action by reducing the value of the home currency and causing export spending by foreign residents to fall. This helps to reduce aggregate demand on net and reduces the price level.
   c. reinforce the main effect of the policy action by reducing the value of the home currency and causing export spending by foreign residents to rise. This helps to raise aggregate demand on net and raises the price level.
   d. reinforce the main effect of the policy action by raising the value of the home currency and causing export spending by foreign residents to fall. This helps to reduce aggregate demand on net and reduces the price level.

8. If the actual unemployment rate equals the natural unemployment rate, then
   a. no inflation is possible.
   b. long-run macroeconomic equilibrium exists.
   c. no unemployment exists.
   d. All of the above

9. If the current unemployment rate is equal to both the natural rate of unemployment and the nonaccelerating inflation rate of unemployment, then
   a. there is pressure for the inflation rate to increase at the current short-run macroeconomic equlibrium.
   b. there is pressure for the inflation rate to decrease at the current short-run macroeconomic equlibrium.
   c. there is a steady inflation rate at a long-run macroeconomic equilibrium.
   d. the inflation rate is presently below its long-run equilibrium level.

10. The natural rate of unemployment equals the actual unemployment rate
    a. in the long run.
    b. only when the actual inflation rate is zero.
    c. only when the natural inflation rate is zero.
    d. only in the short run.

11. The natural rate of unemployment
    a. occurs when the inflation rate is correctly anticipated.
    b. is always above the actual rate of unemployment.
    c. is always below the actual rate of unemployment.
    d. occurs usually in the short run.

12. The only way to keep the actual unemployment rate below the natural unemployment rate is to
    a. have the actual inflation rate be less than the expected inflation rate.
    b. have the actual inflation rate be higher than the expected inflation rate.
    c. constantly reduce the inflation rate.
    d. provide job security.

13. The rational expectations hypothesis
    a. maintains that policymakers cannot get economic agents to make systematic forecasting errors.
    b. maintains that economic agents base their forecasts on their understanding of how the economy operates.
    c. maintains that economic agents will use all information, including expected stabilization policies, when they estimate the future inflation rate.
    d. All of the above

14. The policy irrelevance proposition maintains that
    a. anticipated government stabilization policies cannot reduce unemployment below the natural rate.
    b. unanticipated stabilization policies cannot reduce unemployment below the natural rate.
    c. anticipated stabilization policies cannot reduce the increase in actual unemployment resulting from a decrease in the inflation rate.
    d. unanticipated stabilization policies are very effective in reducing actual unemployment below natural unemployment.

15. Economists who promote the policy irrelevance proposition
    a. reject the rational expectations approach.
    b. assume that price and wages are flexible.
    c. maintain that monetary policy cannot affect real variables in the short run.
    d. established the proposition by building directly on ideas developed by Keynes.

16. When expansionary monetary policy is unanticipated, in the short run
    a. the unemployment rate will fall.
    b. actual inflation will exceed anticipated inflation.
    c. macroeconomic equilibrium will lie to the right of the long-run aggregate supply curve.
    d. All of the above

17. When contractionary monetary policy is fully and correctly anticipated, in the short run
    a. the unemployment rate will fall.
    b. the actual decline in inflation will equal the anticipated decline in inflation.
    c. macroeconomic equilibrium will lie to the left of the long-run aggregate supply curve.
    d. All of the above

18. Other things being equal, a rightward shift in the Phillips curve could be caused by
    a. a decline in the actual inflation rate.
    b. a decline in the rate of unemployment.
    c. an increase in the expected inflation rate.
    d. an increase in the rate of growth of the money supply.

19. The real business cycle approach
    a. assumes that prices and wages are flexible.
    b. maintains that real economic variables can change even if mistakes are not made.
    c. suggests that supply creates its own demand.
    d. All of the above

20. The efficiency wage theory best provides an explanation for
    a. inflation.
    b. sticky prices.
    c. unemployment.
    d. flexible real wages.

MATCHING
Choose the item in column (2) that best matches an item in column (1).

	(1)		(2)
a.	Phillips curve	g.	vertical LRAS curve
b.	NAIRU	h.	efficiency wage hypothesis
c.	real-wage rigidity	i.	noninflationary unemployment
d.	long-run Phillips curve	j.	supply creates its own demand
e	policy irrelevance proposition	k.	rational expectations hypothesis
f.	real business cycle approach	l.	relationship between inflation and unemployment

WORKING WITH GRAPHS

1. Suppose you are given the Phillips curve in the graph below.

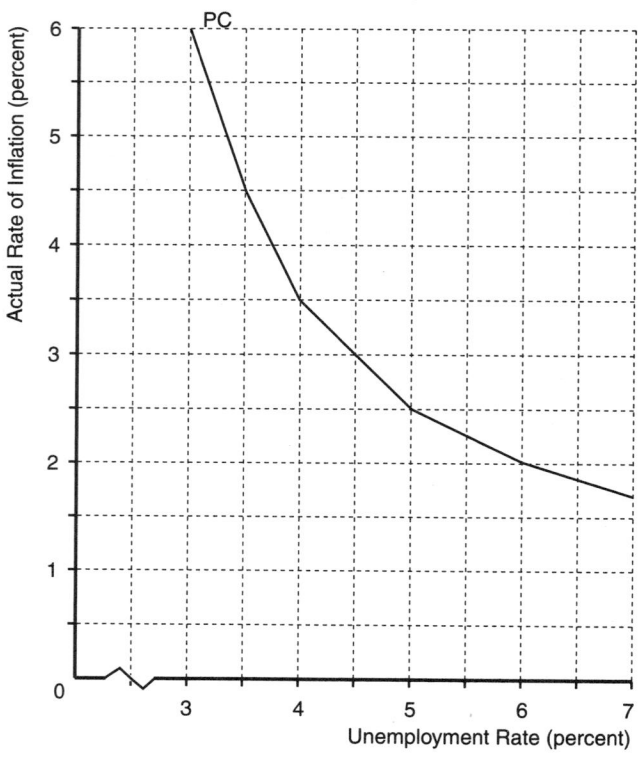

   a. If an unemployment rate of 4 1/2 percent constitutes "full" employment, what rate of inflation should the economy expect at full employment?
   b. Suppose the government were to set a goal of lowering inflation to 2 percent. Assuming the Phillips curve above is accurate and stable, what rate of unemployment will have to be tolerated if the goal for inflation is to be achieved?
   c. Suppose the Phillips curve above describes the inflation-unemployment trade-off with individuals anticipating inflation of 3 percent. If individuals suddenly begin to expect inflation of 4 1/2 percent, and this raises the rate of unemployment for each level of actual inflation by 1 percent, draw the new Phillips curve.
   d. What will be the rate of unemployment if the actual rate of inflation is the anticipated rate, 4 1/2 percent?
   e. After the rise in expected inflation, if the actual rate of inflation turns out to be 3 percent rather than 4 1/2 percent, what will the rate of unemployment be?
   f. What conclusion can be drawn from your answers to parts c through e?

2. Analyze the graphs below, then answer the questions that follow.

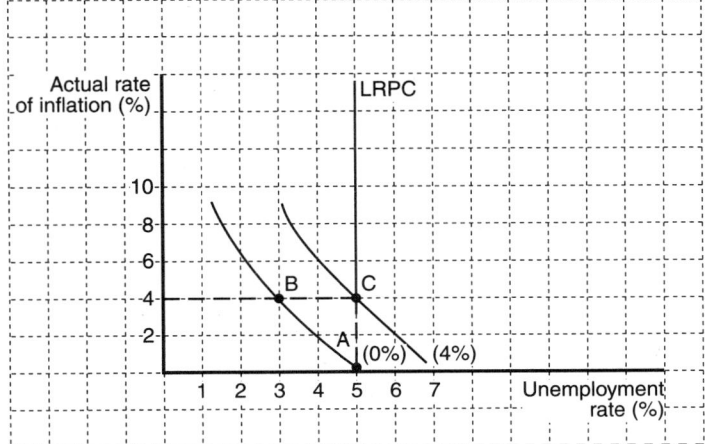

a. What is the natural rate of unemployment for the economy depicted in the graphs above?
b. Starting from point A and assuming that economic agents anticipated 0 percent inflation, what will be the actual unemployment rate if the actual inflation rate is 2 percent? 4 percent?
c. If the anticipated inflation rate is 4 percent, and the actual inflation rate is 4 percent, what will be the actual unemployment rate? The natural unemployment rate?
d. When does the short-run Phillips curve shift?

PROBLEMS

1. In the table below are three short-run Phillips curves. Columns 1 and 2 show the relationship between the unemployment rate (U) and the actual rate of inflation (R) when the anticipated rate of inflation in the economy is 0 percent; columns 3 and 4 when the anticipated R is 6 percent; and columns 5 and 6 when the anticipated R is 12 percent.

Phillips Curve 1		Phillips Curve 2		Phillips Curve 3	
(1) U	(2) R_0%	(3) U	(4) R_6%	(5) U	(6) R_{12}%
10	0	10	6	10	12
9	1	9	7	9	13
8	3	8	9	8	15
7	6	7	12	7	18
6	10	6	16	6	22
5	15	5	21	5	27
4	21	4	27	4	33

a. Suppose the anticipated rate of inflation and the actual rate of inflation are 0. The unemployment rate U is _____ percent.

b. If expansionary monetary or fiscal policies are used to increase aggregate demand and to reduce U to 7 percent, the actual R will (increase/decrease) to _____ percent.

c. When the economy comes to anticipate this R, U will be _____ percent; and if expansionary stabilization policies are again used to reduce U to 7 percent, the actual R will to _____ percent.

d. And when the economy comes to anticipate this R, U will be _____ percent; if stabilization policies are once more used to reduce U to 7 percent, the actual R will _____ to _____ percent; and when the economy comes to anticipate this R, U will again _____ to _____ percent.

e. Use your answers to parts (a), (b), (c), and (d) above to show U in the long run at each of the actual R's in the table.

R	U
0	____
6	____
12	____

ANSWERS TO CHAPTER 17

COMPLETION QUESTIONS

1. inflation; unemployment
2. is not
3. long; equilibrium; equals; is not
4. greater than; accelerating
5. cannot; little; will; will not
6. rise; overestimate; will not
7. pure competition; price/wage flexibility; policymakers; economic agents
8. no; will; is not; will not
9. rational; are not
10. high wages lead to high productivity; inflexible
11. flexible; supply; oil prices; technology; composition of the labor force

TRUE-FALSE QUESTIONS

1. F It relates inflation rates to unemployment rates.
2. T
3. T
4. F The NAIRU is the unemployment rate consistent with steady inflation, which can potentially adjust during the course of cyclical adjustments in the economy, whereas the natural unemployment rate depends on structural factors in the labor market and thereby tends to change over relatively lengthy intervals.
5. T
6. T
7. F It fools people only in the short run.
8. F Unemployment will not be affected very much, if policy is announced.
9. T
10. F In long-run equilibrium they are equal.
11. T
12. F Even if people use both past and current information and their own understanding of how the economy operates, they may still fail to fully anticipate changes in the inflation rate.
13. T
14. F They can change unemployment, but the direction is uncertain.
15. F If the idea of real business cycles is correct, then the ability of activist policies to influence real variables such as employment and real GDP is greatly limited.
16. F The efficiency wage theory indicates that firms will tend to maintain steady real wages over time even as new workers enter the labor force, so unemployment can be long-lasting.
17. T

MULTIPLE CHOICE QUESTIONS

1.c; 2.b; 3.b; 4.a; 5.a; 6.a; 7.d; 8.b; 9.c; 10.a;
11.a; 12.b; 13.d; 14.a; 15.b; 16.d; 17.b; 18.c; 19.d; 20.c.

MATCHING

a and l; b and i; c and h; d and g; e and k; f and j

WORKING WITH GRAPHS

1. a. 3 percent
   b. 6 percent
   c. see graph below

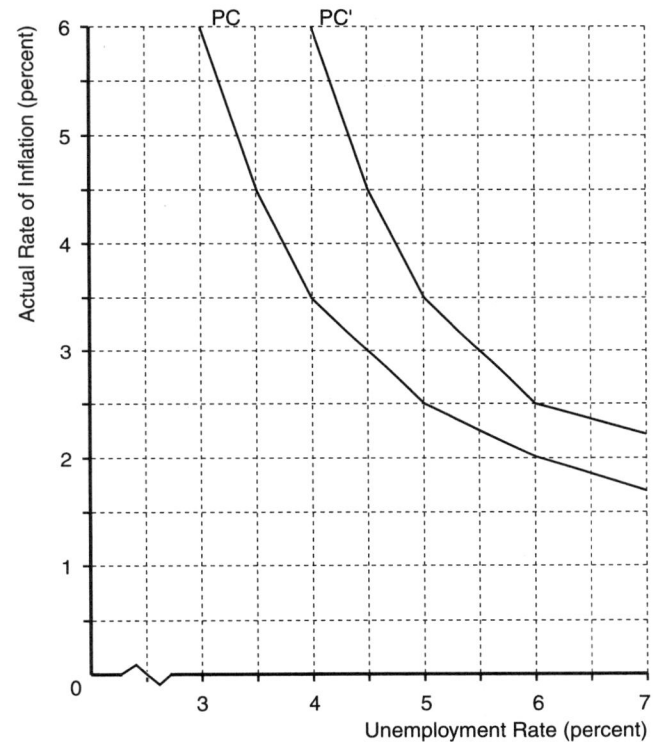

   d. 4 1/2 percent
   e. 5 1/2 percent
   f. When actual inflation is less than anticipated inflation, the unemployment rate increases.

2. a. 5 percent
   b. 4 percent; 3 percent
   c. 5 percent; 5 percent
   d. Every time the anticipated inflation rate changes.

PROBLEMS

1. a. 10
   b. increase; 6
   c. 10; increase; 12
   d. 10; increase; 18; increase; 10
   e. 10 percent; 10 percent; 10 percent

GLOSSARY TO CHAPTER 17

Active (discretionary) policymaking All actions on the part of monetary and fiscal policymakers that are undertaken in response to or in anticipation of some change in the overall economy.

Efficiency wage The optimal wage that firms must pay to maintain worker productivity.

Nonaccelerating inflation rate of unemployment (NAIRU) The rate of unemployment below which the rate of inflation tends to rise and above which the rate of inflation tends to fall.

Passive (nondiscretionary) policymaking Policymaking that is carried out in response to a rule. It is therefore not in response to an actual or potential change in overall economic activity.

Phillips curve A curve showing the relationship between unemployment and changes in wages or prices. It was long thought to reflect a trade-off between unemployment and inflation.

Policy irrelevance proposition The conclusion that policy actions have no real effects in the short run if the policy actions were anticipated and none in the long run even if the policy actions were unanticipated.

Rational expectations hypothesis A theory stating that people combine the effects of past policy changes on important economic variables with their own judgment about the future effects of current and future policy changes.

Small menu costs Costs that deter firms from changing prices in response to demand changes—for example, costs of renegotiating contracts or printing new price lists.

CHAPTER 18
POLICIES AND PROSPECTS FOR GLOBAL ECONOMIC GROWTH

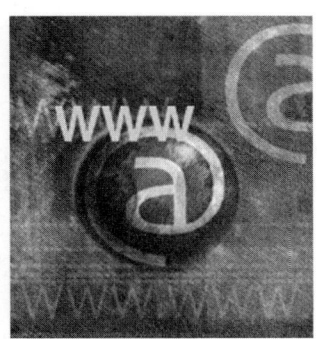

LEARNING OBJECTIVES
After you have studied this chapter you should be able to

1. define dead capital, economic freedom, foreign direct investment, international financial crisis, International Monetary Fund, portfolio investment, quota subscription, and World Bank;

2. explain why population growth can have indefinite effects on a nation's economic growth;

3. discuss how the existence of dead capital hinders investment and economic growth in large portions of the developing world;

4. describe how government inefficiencies have contributed to the creation of large amounts of dead capital in the world's developing nations;

5. identify the rationales for foreign financing of investment in developing nations;

6. explain how developing countries benefit from international capital investment;

7. list the functions of the World Bank and the International Monetary Fund;

8. describe recent criticisms of policymaking at the World Bank and the International Monetary Fund.

CHAPTER OUTLINE

1. Population has ambiguous effects, in principle, on growth in any nation's per capita income: A rise in population with aggregate real GDP unchanged reduces per capita real income, but if the

population increase translates into increased labor force participation and production of goods and services, on net it can boost per capita real income.
- a. The rate of growth of a nation's per capita real GDP equals the rate of growth in its real GDP minus the rate of growth of its population.
- b. In some nations, such as China, India, and Pakistan, positive economic growth has been associated with population growth, but in others, such as Saudi Arabia, Niger, and Zambia, per capita real incomes have declined as populations have increased.

2. Human freedoms are related to economic growth, but not all forms of freedom contribute equally to growth.
- a. Greater economic freedom, such as having increased rights to own and exchange private property, unambiguously adds to economic growth.
- b. Increased political freedom has less clear-cut effects on economic growth; greater economic freedom appears to boost economic growth, which in turn often leads to more political freedom.

3. A significant factor hindering growth in many nations is that many of their productive resources are dead capital, which are capital goods lacking clear title of ownership.
- a. Governments in many nations place numerous legal restraints on the ability to exchange capital resources, and many of these governments inefficiently enforce regulations, and this further obstructs allocation of capital resources to their most efficient uses.
- b. People often resort to bribery and other forms of corruption to overcome government regulations and inefficiencies, but corruption makes production more costly and thereby hampers economic growth.

4. Financial intermediation is not well developed in many nations, which contributes to their reliance on flows of funds from abroad to finance significant shares of their overall capital investment.
- a. Foreign direct investment entails a purchase of at least a 10 percent share of ownership of business resources abroad, while portfolio investment involves a purchase of less than a 10 percent share.
- b. Asymmetric information problems, such as adverse selection and moral hazard, can particularly hinder capital accumulation and growth in developing nations. In many cases, foreign investors who can withdraw funds quickly may have an advantage over domestic investors. This helps explain why a number of developing nations rely on about $100 billion per year in international funds to help finance internal investment.
- c. Because many foreign investors have the ability to shift funds among nations speedily, there is always a potential for a country that makes use of significant flows of international funds to experience an international financial crisis.

5. The World Bank is a multinational agency that specializes in making loans intended to promote long-term development and growth in about 100 developing nations.
- a. The World Bank actually encompasses five different institutions, which specialize in various aspects of assisting developing nations through loans and investment dispute settlement.
- b. The World Bank extends about $20 billion in loans to developing countries using funds it receives from governments of the world's wealthiest nations and funds it raises in private financial markets.

6. The International Monetary Fund (IMF) is a multinational organization that aims to promote world economic growth through greater financial stability.
- a. Nations joining the IMF deposit funds in an account called the country's quota subscription, which determines the share of each nation's voting rights and its borrowing limits; funds are denominated in an international unit of accounting called special drawing rights.

b. Originally the IMF offered short-term loans, but since the 1970s the IMF has also provided longer-term loans to poor and heavily indebted nations.

KEY TERMS
Dead capital
Economic freedom
Foreign direct investment
International financial crisis
International Monetary Fund
Portfolio investment
Quota subscription
World Bank

KEY CONCEPTS
Averse selection
Asymmetric information
Corruption
Government inefficiency
Moral hazard

COMPLETION QUESTIONS
Fill in the blank or circle the correct term.

1. The rate of growth in per capita real GDP equals the rate of growth of _____ minus the rate of growth of the nation's _____.

2. There is considerable evidence that greater _____ freedom contributes to economic growth, but increased _____ freedom does not always appear to promote greater economic growth.

3. Equipment and machines that cannot be directed to their most efficient uses because owners cannot establish legal title of ownership are examples of _____.

4. Greater government inefficiency tends to be associated with (lower; higher) economic growth.

5. When an individual purchases a 20 percent share of ownership in the productive resources of a business located in another nation, this is an example of _____.

6. When an individual purchases a 1 percent share of ownership in the productive resources of a business located in another nation, this is an example of _____.

7. Today, _____ is the most common means by which international investors provide funds to finance capital accumulation in developing nations, and _____ is the least common source of such funds from international investors..

8. The World Bank is actually composed of (three, five, seven) different institutions with between 137 and 182 member nations.

9. When a nation joins the International Monetary Fund, its government must establish a _____ by depositing funds into a special account, and the amount it deposits determines its share of _____ rights and how much it can apply to _____ from the IMF.

10. A nation's quota subscription in the International Monetary Fund is denominated in an international accounting unit known as _____.

11. The rapid withdrawal of foreign investments and loans from a nation is called _____.

TRUE-FALSE QUESTIONS
Circle the **T** if the statement is true and the **F** if it is false. Explain to yourself why a statement is false.

T F 1. Estimates indicate that the total amount of real estate not legally registered in developing countries is nearly equal to the total market value of all companies listed on stock exchanges in developed nations.

T F 2. If a country's population growth rate exceeds the rate of growth of its total output of goods and services, then this country experiences positive economic growth.

T F 3. Nations containing less than 20 percent of the world's people produce more than 80 percent of global output of goods and services.

T F 4. Based on worldwide evidence, economic growth is positively related to the level of corruption.

T F 5. In most developing nations, ownership of dead capital is readily transferred from one individual to another.

T F 6. Since the 1970s, bank loans have accounted for a steadily increasing share of international funding for investment projects in developing nations.

T F 7. In the 1970s and early 1980s, less than 10 percent of all international financial flows to developing nations was in the form of portfolio investment, but now portfolio investment accounts for more than 40 percent of these flows of funds.

T F 8. An example of a moral hazard problem that can hamper investment and growth in a developing nation is the inability of a bank located in that country to monitor how businesses use funds they have borrowed from the bank.

T F 9. An example of an adverse selection problem that can hamper investment and growth in a developing nation is the inability of a wealthy individual to determine whether a businessperson who has asked for a loan to buy capital equipment can maintain legal ownership of the equipment once it has been purchased.

T F 10. The World Bank's main duty is to provide temporary credit to the government of a nation experiencing major short-term financial problems.

T F 11. The United States is the nation with the largest share of voting rights in the International Monetary Fund.

T F 12. The International Monetary Fund makes only loans that must be repaid within three months.

CHAPTER 18: POLICIES AND PROSPECTS FOR GLOBAL ECONOMIC GROWTH

MULTIPLE CHOICE QUESTIONS
Circle the letter that corresponds to the best answer.

1. During the past decade, a nation's average annual rate of population growth was 4 percent, and its average annual real GDP growth rate was 1 percent. Thus, the average annual rate of growth in per capita GDP was
   a. -3.0 percent.
   b. -4.0 percent.
   c. +3.0 percent.
   d. +0.25 percent.

2. Between 1984 and 2003, a nation's population grew by 3 percent every year. From 1984 to 1993, its real GDP increased at a rate of 2 percent each year. From 1994 to 2003, however, productivity improvements increased the rate of growth of real GDP to 4 percent per year. It may be concluded that this country experienced
   a. positive economic growth from 1984 to 1993 and negative economic growth from 1994 to 2003.
   b. negative economic growth from 1984 to 1993 and positive economic growth from 1994 to 2003.
   c. a 5 percent increase in per capita real GDP from 1984 to 1993 and a 7 percent increase in per capita real GDP from 1994 to 2003.
   d. a 1 percent increase in per capita real GDP from 1984 to 1993 and a 2 percent increase in per capita real GDP from 1994 to 2003.

3. An increase in a nation's population
   a. increases the labor force and thus the production of goods and services and thereby unambiguously raises the nation's per capita income.
   b. distributes output of goods and services over a larger number of people and thereby unambiguously reduces the nation's per capita income.
   c. decreases labor force participation and thus reduces total output of goods and services, which also must be distributed over a larger number of people, thereby unambiguously reducing the nation's per capita income.
   d. increases the labor force and thus raises total output of goods and services, which must be distributed over a larger number of people, thereby having potentially offsetting effects on the nation's per capita income.

4. Which of the following statements concerning the relationship between human freedom and economic growth is correct?
   a. Increased political freedom does more than greater economic freedom to boost a typical nation's economic growth.
   b. Increased economic freedom does more than greater political freedom to boost a typical nation's economic growth.
   c. Greater economic freedom can only contribute to economic growth if accompanied by increased political freedom.
   d. Greater political freedoms, such as freedom of speech and the press, unambiguously contribute to lower economic growth.

5. Capital resources that lack clear title of ownership are known as
   a. unsubscribed capital.
   b. nonfinancial capital.
   c. indirect capital.
   d. dead capital.

6. A key factor contributing to the dead capital problem in many nations is
   a. inefficiencies in government enforcement of ownership rights to resources.
   b. numerous bureaucratic rules for establishing legal ownership of resources.
   c. the inability to establish clear legal title of ownership of resources.
   d. All of the above.

7. Which of the following currently accounts for the largest percentage of international flows of funds to developing nations?
   a. foreign direct investment
   b. asymmetric investment
   c. portfolio investment
   d. foreign bank loans

8. When an individual U.S. investor purchases more than 10 percent of the shares of a business in a developing nation, she has engaged in
   a. foreign direct investment.
   b. asymmetric investment.
   c. portfolio investment.
   d. a foreign bank loan.

9. One way that the World Bank might experience an adverse selection in its own lending is if
   a. a corrupt government leadership applies for a World Bank loan with the stated aim of using them for capital investment but with the true intention of squandering all the funds.
   b. an inefficient government that has already obtained a World Bank loan disburses the funds to private investors who are unable to establish legal ownership of capital resources.
   c. it has made a loan to a private company that has decided to redirect the funds to a riskier project than the capital investment the funds were originally intended to finance.
   d. it has made a loan to a private company that hires incompetent contractors to construct and equip a new factory.

10. The main difference in the International Monetary Fund's functions today as compared with its functions when it was established is that it now
    a. makes short-term loans, whereas originally it mainly made only long-term loans.
    b. makes long-term loans, whereas originally it mainly made only short-term loans.
    c. is concerned only with promoting long-term growth, whereas originally its focus was solely on assisting with short-term adjustments.
    d. is concerned only with assisting with short-term adjustments, whereas originally its focus was solely on promoting long-term growth.

11. The largest share of World Bank loans goes to nations located in
    a. eastern and southern Asia.
    b. the Middle East.
    c. Latin America.
    d. Africa.

12. The World Bank's primary focus is on
    a. providing short-term loans to governments of nations to head off international financial crises before they occur.
    b. providing short-term loans to governments of nations that are already experiencing international financial crises.
    c. forecasting international financial crises that may adversely affect investment and growth in developing nations.
    d. providing long-term loans intended to promote investment and growth in many of the world's poorest developing nations.

CHAPTER 18: POLICIES AND PROSPECTS FOR GLOBAL ECONOMIC GROWTH

13. In the poorest 25 percent of the world's nations, obtaining legal authorization to operate a business entails incurring an average expense equal to
    a. 8 percent of a typical resident's annual earnings, or about one month of earnings.
    b. 60 percent of a typical resident's annual earnings, or about seven months of earnings.
    c. 75 percent of a typical resident's annual earnings, or about nine months of earnings.
    d. 115 percent of a typical resident's annual earnings, or about fourteen months of earnings.

14. In the countries with the highest per capita real incomes, typically _____ legal steps must be completed to establish a business, but in the countries with the lowest per capita incomes, typically _____ legal steps are required.
    a. at least 16; no more than 5.
    b. no more than 5; at least 16.
    c. zero; at least 16.
    d. at least 16; zero.

15. Critics argue that the World Bank and International Monetary Fund may actually make international financial crises more likely by
    a. encouraging governments in developing nations to engage in more risky practices in the knowledge that these multinational institutions are likely to provide financial assistance.
    b. encouraging banks toughen to credit standards for loans to developing nations in light of fears that these multinational institutions will withhold financial assistance.
    c. offering subsidized credit to developing nations and thereby improving the ability of private investors to assess which nations are most creditworthy.
    d. offering credit under very tough initial conditions and loosening the conditions after a nation has demonstrated its ability to put funds to wise use.

16. A key change in IMF policymaking that the International Monetary Fund has itself proposed is to
    a. replace the IMF and World Bank with a single institution oriented solely toward short-term lending to nations experiencing temporary financial problems.
    b. replace the IMF and World Bank with a single institution that would conduct open market operations like a central bank.
    c. develop better systems for predicting financial crises to raise the likelihood of heading them off before they occur.
    d. replace the current management of the IMF with chief financial officials of member nations of the organization.

MATCHING
Choose the item in column (2) that best matches an item in column (1).

(1)	(2)
a. growth rate of per capita real GDP	g. international financial crisis
b. World Bank	h. quota subscription
c. IMF member shares	i. long-term loans to developing nations
d. sudden cutoff in funds to a nation	j. dead capital
e. foreign direct investment	k. purchase of more than 10 percent of a business located in another country
f. productive equipment that legal restraints prevent from being used efficiently	l. real GDP growth rate minus population growth rate

PROBLEMS

1. In the table are annual rates of growth of real GDP and population growth rates for countries A, B, and C, with all amounts expressed as a percentage rate of growth per year.

Year	Country A		Country B		Country C	
	Real GDP Growth	Population Growth	Real GDP Growth	Population Growth	Real GDP Growth	Population Growth
2001	4	4	3	3	1	3
2002	4	4	3	3	3	4
2003	4	3	3	3	8	5
2004	4	2	3	3	9	7
2005	4	2	3	3	11	8

   a. During the interval 2001-2005, the average annual rate of economic growth in country A was _____ percent.

   b. During the interval 2001-2005, the average annual rate of economic growth in country B was _____ percent.

   c. During the interval 2001-2005, the average annual rate of economic growth in country C was _____ percent.

   d. How do your answers to parts a, b, and c help to illustrate why *both* aggregate real GDP growth *and* population growth must be taken into account in assessing a nation's economic growth?

ANSWERS TO CHAPTER 18

COMPLETION QUESTIONS

1. aggregate real GDP; population
2. economic; political
3. dead capital
4. lower
5. foreign direct investment
6. portfolio investment
7. portfolio investment; foreign direct investment
8. five
9. quota subscription; voting; borrow
10. special drawing rights
11. an international financial crisis

TRUE-FALSE QUESTIONS

1. T
2. F   In this situation, the rate of growth of per capita real GDP is negative.
3. T
4. F   There is a negative relationship between the level of corruption and economic growth.
5. F   By definition, lack of clear title of ownership makes dead capital hard to exchange.
6. F   In fact, bank loans have declined as a percentage of total international investment funds.
7. T
8. T
9. T
10. F The World Bank concentrates on making long-term loans intended to help finance a nation's development and growth.
11. T
12. F Originally the IMF did specialize in short-term lending, but today it makes both short-term and long-term loans.

MULTIPLE CHOICE QUESTIONS

1.a; 2.b; 3.d; 4.b; 5.d; 6.d; 7.c; 8.a; 9.a; 10.b;
11.a; 12.d; 13.d; 14.b; 15.a; 16.c

MATCHING

a and l;   b and i;   c and h;   d and g;   e and k;   f and j

PROBLEM

1. a. The answer is the average of 0% growth in 2001, 0% growth in 2002, 1% growth in 2003, 2% growth in 2004, and 2% growth in 2005, or (0% + 0% + 1% + 2% + 2%)/5 = 1%.
   b. The answer is the average of 0 percent growth in each year, or 0%.
   c. The answer is the average of -2% growth in 2001, -1% growth in 2002, 3% growth in 2003, 2% growth in 2004, and 3% growth in 2005, or ((-2%) + (-1%) + 3% + 2% + 3%)/5 = 1%.
   d. In countries A and C, average growth in real GDP per capita was the same over the period even though population growth declined somewhat in country A and rose in country C. Over the entire period, the growth of aggregate real GDP in country C rose more than sufficiently relative to the increase in the nation's population growth to keep average annual growth the same as in country A, at 1 percent per year. In country B, aggregate real GDP growth was the same as population growth every year, so real per capital GDP growth remained equal to 0 percent throughout the period. Thus, country B's people were no better off, on average, in 2005 than they were in 2001.

GLOSSARY TO CHAPTER 18

Dead capital Any capital resource that lacks clear title of ownership.

Economic freedom The rights to own private property and to exchange goods, services, and financial assets with minimal government interference.

Foreign direct investment The acquisition of more than 10 percent of the shares of ownership in a company in another nation.

International financial crisis The rapid withdrawal of foreign investments and loans from a nation.

International Monetary Fund A multinational organization that aims to promote world economic growth through more financial stability.

Portfolio investment The purchase of less than 10 percent of the shares of ownership in a company in another nation.

Quota subscription A nation's account with the International Monetary Fund, denominated in special drawing rights.

World Bank A multinational agency that specializes in making loans to about 100 developing nations in an effort to promote their long-term development and growth.

CHAPTER 32
COMPARATIVE ADVANTAGE AND THE OPEN ECONOMY

LEARNING OBJECTIVES

After you have studied this chapter, you should be able to

1. define absolute advantage, comparative advantage, quota system, dumping, infant industry argument, General Agreement on Tariffs and Trade (GATT), World Trade Organization (WTO), voluntary import expansion (VIE), and voluntary restraint agreement (VRA);

2. explain how nation pays for its imports;

3. distinguish between absolute advantage and comparative advantage and determine each for nations, given sufficient information;

4. interpret a graph that shows the effect of an import quota;

5. interpret a graph that shows the effect of a tariff;

6. relate comparative advantage to opportunity cost;

7. list reasons why trade arises between nations;

8. list potential costs of international trade;

9. identify the pros and cons of arguments against free trade among nations;

10. contrast and compare the results of import quotas and tariffs.

CHAPTER OUTLINE

1. The proportion of GDP accounted for by trade varies greatly among individual nations, but if trade were curtailed, even such nations as the United States would be affected significantly.

214 CHAPTER 32: COMPARATIVE ADVANTAGE AND THE OPEN ECONOMY

    a. A nation ultimately pays for its imports by exports; thus restrictions on imports ultimately reduce exports.
    b. If trade is voluntary, then both nations participating in an exchange benefit.

2. A nation has an absolute advantage in the production of good A if it can produce more units of good A than other nations can, from a given quantity of inputs; a nation has a comparative advantage in producing good A if out of all the goods it can produce, good A has the lowest opportunity cost.

3. Opportunity costs, and hence comparative advantages, differ among nations.

4. There are numerous arguments that have been presented as being anti–free trade. Such arguments include the infant industry argument, protecting domestic jobs, countering foreign subsidies and dumping, protecting the environment, and national defense concerns. Some of these arguments are simply wrong, and others emphasize costs to the neglect of benefits.

5. There are numerous methods that nations have used to restrict foreign trade.
    a. Some nations place import quotas on foreign goods.
    b. Some nations place taxes or tariffs on foreign goods.
    c. Both quotas and tariffs raise prices to domestic consumers and reduce the quantity of goods traded.

KEY TERMS
General Agreement on Tariffs and Trade (GATT)
Infant industry argument
Quota system
World Trade Organization (WTO)

KEY CONCEPTS
Comparative advantage
Absolute advantage
Dumping
Voluntary import expansion (VIE)
Voluntary restraint agreement (VRA)

COMPLETION QUESTIONS
Fill in the blank, or circle the correct term.

1. A nation ultimately pays for imports by _____.

2. If world trade ceased to exist, all trade-related jobs (would, would not) be lost in the long run; instead nations would simply _____. Nevertheless, worldwide living standards would (fall, rise) significantly.

3. International trade permits each nation to specialize in the production of those goods for which it has a(n) _____ advantage; each nation specializes in the production of goods for which its opportunity costs are the (lowest, highest).

4. Nations have an incentive to specialize and trade because they have different collective tastes and because different nations will always have different _____ costs to producing goods.

5. There are numerous arguments against free trade; they include the _____ industry argument, the argument that trade may impede _____ protection, and the contention that trade may interfere with national _____ efforts.

6. Two ways to restrict foreign trade analyzed in the text are _____ on imports and _____ on imported goods.

7. Most restrictions on international trade have one major element in common: they interfere with nations' specializing in the production of goods for which they have a(n) _____ advantage. Therefore they are economically (inefficient, efficient).

8. Because a nation ultimately pays for imports with its _____, restricting imports to save jobs destroys jobs in the _____ sector of the economy; hence, on net, import restrictions (do, do not) save jobs.

TRUE-FALSE QUESTIONS
Circle the **T** if the statement is true, the **F** if it is false. Explain to yourself why a statement is false.

T F 1. If all world trade ceased, import sector jobs and export sector jobs would be permanently destroyed.

T F 2. Because international trade is voluntary in the private sector, both nations benefit from trade that is continued.

T F 3. Imports are paid for by exports.

T F 4. In effect, a tariff makes the supply of the good in question a vertical line at a level below the original equilibrium quantity.

T F 5. A U.S. tariff on Japanese-made goods will lead to an increase in the demand for U.S. goods that are substitutes for those Japanese-made goods.

T F 6. Import quotas harm domestic consumers but help domestic producers of those goods on which quotas are placed.

T F 7. Tariffs harm domestic consumers and harm domestic producers of goods that compete with the goods on which tariffs are placed.

T F 8. A tariff on good X will cause a leftward shift of the supply curve for good X in the foreign country, and a rightward shift of the demand curve for good X in the country that imposed the tariff.

T F 9. In a two-country world, it is possible for both countries to have a comparative advantage in the production of a specific good.

T F 10. If the United States has a comparative advantage in producing wheat, it must be true that the opportunity cost for producing wheat in the United States is below that opportunity cost in other nations.

T F 11. Because in the real world nations have different resource endowments and different collective tastes, trade will always be advantageous.

T F 12. It is easy to determine the industries to which the infant industry argument applies.

T F 13. If a nation imposes anti-dumping laws, its consumers will pay lower prices for goods.

T F 14. Free trade may increase a nation's instability in the short run, because over time a nation's comparative advantage can change.

T F 15. When a nation restricts imports to protect jobs, it in effect preserves less productive employment at the expense of more productive employment.

T F 16. One difference between the economic effects of quotas versus tariffs is that tariffs lead to a higher price to consumers but quotas do not.

MULTIPLE CHOICE QUESTIONS

Circle the letter that corresponds to the best answer.

1. In the long run, a nation pays for its imports by
   a. exporting.
   b. creating money.
   c. extending credit to the exporting nation.
   d. All of the above

2. The U.S. ratio of imports to GDP is about _____ percent.
   a. 5
   b. 10
   c. 15
   d. 20

3. If trade between two nations is voluntary and continued, then
   a. both nations benefit.
   b. one nation could benefit more than the other.
   c. living standards are higher in both nations than if trade were not permitted.
   d. All of the above

4. Country A can produce both wheat and oranges using fewer resources than country B. Which of the following statements is true?
   a. Country A has a comparative advantage in producing both goods.
   b. Country A has an absolute advantage in producing both goods.
   c. Country B has no comparative advantage.
   d. Country B must have an absolute advantage in producing one of the goods.

5. If Country C has a comparative advantage in producing wheat, then its opportunity cost of producing wheat
   a. is maximized.
   b. equals the opportunity cost of producing other goods.
   c. cannot be determined.
   d. is lowest among its trading partners.

6. Nations find it advantageous to trade because they
   a. have different resource endowments.
   b. have different collective tastes.
   c. have different comparative advantages.
   d. All of the above

CHAPTER 32: COMPARATIVE ADVANTAGE AND THE OPEN ECONOMY

7. Which of the following is **NOT** an argument used against free trade?
   a. Free trade makes nations more interdependent.
   b. Free trade leads nations to specialize in production.
   c. Free trade increases average and total worldwide incomes.
   d. Imports may destroy some domestic jobs.

8. Which of the following is most **UNLIKE** the others?
   a. import quota
   b. tariff
   c. free trade
   d. anti-dumping laws

9. Concerning import quotas and tariffs, which of the following statements is true?
   a. Both lead to lower prices for consumers.
   b. Both lead to more imports.
   c. Tariffs lead to higher prices, but quotas do not.
   d. Quotas directly restrain imports, while tariffs induce people to choose fewer imports.

10. Which statement is **NOT** true, concerning the use of import restrictions to save jobs?
    a. The cost to consumers often exceeds the value of the jobs saved.
    b. Some jobs are destroyed in the export sector.
    c. In the long run they do not save jobs in those industries in which a nation has lost its comparative advantage.
    d. They are the most efficient way to help domestic workers threatened by foreign competition.

MATCHING
Choose the item in column (2) that best matches an item in column (1).

(1)	(2)
a. anti-dumping law	e. minimum opportunity cost of production
b. tariff	f. trade restriction
c. comparative advantage	g. tax on foreign-produced goods
d. absolute advantage	h. producing at a lower cost

WORKING WITH GRAPHS

1. Analyze the graphs below, then answer the questions that follow. They deal with an import quota set on foreign-made sugar. Start at point A.

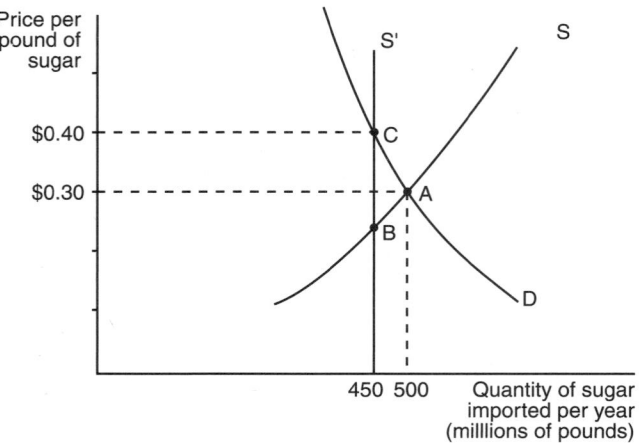

   a. What is the price of sugar in this country, without the import quota?
   b. What is the maximum amount of imported sugar given the quota?
   c. What is the price of sugar, given the import quota?
   d. What is the effective import supply curve, given the quota?

2. Analyze the graphs below, which deal with a U.S. tariff placed on Japanese-made autos, then answer the questions that follow. Start with point A.

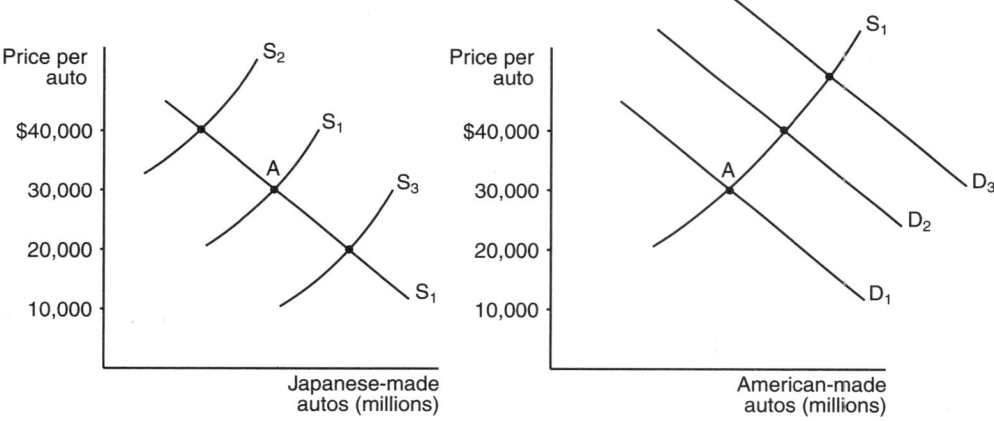

   a. What is the price of U.S.- and Japanese-made autos in the U.S., without the tariff?
   b. Which curve represents the supply of Japanese-made autos into the U.S., after the tariff?
   c. Which curve shows the demand for U.S.-made autos in the U.S., after the tariff?
   d. What is the price of autos in the United States, after the tariff is imposed?

CHAPTER 32: COMPARATIVE ADVANTAGE AND THE OPEN ECONOMY

PROBLEMS

1. Suppose that Germany and the United States are both experiencing full employment and can produce the following amounts of wine and beer per week. Use this information to answer the following questions.

	Wine (gallons)	Beer (gallons)
Germany	600	1200
United States	400	1600

   a. Germany has a comparative advantage in the production of _____, whereas the United States has a comparative advantage in the production of _____.

   b. What is the cost of wine in terms of beer in Germany? What is the cost of wine in terms of beer in the United States?

2. Let us assume we are again in a two-country world, with the countries being Germany and the United States. Again, to simplify, let us assume that there are only two goods produced, coal and steel. In the table that follows, you will find the production possibilities of both countries. Assume that each country is currently operating at combination B on its production possibilities schedule. Use this information to answer the following questions. (Hint: Remember that movements along production possibilities curves involve opportunity costs and that comparative advantage and trade depend on opportunity costs.) Entries are in thousands of tons per week.

		A	B	C	D
Germany:	Coal	0	24	48	72
	Steel	18	12	6	0
United States:	Coal	0	36	72	108
	Steel	36	24	12	0

   a. Which country has an absolute advantage in the production of coal? Steel?

   b. Which country has a comparative advantage in the production of coal? Steel?

   c. What is the cost of coal in terms of steel in Germany? What is the cost of coal in terms of steel in the United States?

   (Remember that these are opportunity costs as determined by production possibilities.)

   d. What is the current world production of coal and steel?
   e. If both countries specialize in the production of goods in which they have a comparative advantage, Germany will produce _____ thousand tons of _____, the United States will produce _____ thousand tons of _____, and the world output of coal will increase by _____ thousand tons.

ANSWERS TO CHAPTER 32

COMPLETION QUESTIONS

1. exporting
2. would not; produce the goods themselves; fall
3. comparative; lowest
4. opportunity
5. infant; environmental; defense
6. quotas; tariffs
7. comparative; inefficient
8. exports; export; do not

TRUE-FALSE QUESTIONS

1. F Eventually each nation will produce its own goods—but at a higher cost and (perhaps) lower quality.
2. T
3. T
4. F That describes the effect of an import quota.
5. T
6. T
7. F They help domestic producers of such goods.
8. T
9. F Not for a *specific* good.
10. T
11. T
12. F In practice it is difficult to predict which industries will eventually be successful without aid.
13. F Consumers will pay higher prices.
14. T
15. T
16. F Both lead to higher prices for consumers.

MULTIPLE CHOICE QUESTIONS

1.a;   2.c;   3.d;   4.b;   5.d;   6.d;   7.c;   8.c;   9.d;   10.d.

MATCHING

a and f;   b and g;   c and e;   d and h

WORKING WITH GRAPHS

1. a. 30 cents per pound
   b. 450 million pounds
   c. 40 cents per pound
   d. The same as the regular supply curve up to 450 million pounds; vertical from that point on.

2. a. $30,000 per auto
   b. S_2
   c. D_2
   d. $40,000 per auto

CHAPTER 32: COMPARATIVE ADVANTAGE AND THE OPEN ECONOMY

PROBLEMS

1. a. wine; beer
    b. 1 gallon wine = 2 gallons beer; 1 gallon wine = 4 gallons beer
2. a. United States; United States
    b. Germany; United States
    c. 4000 tons coal = 1000 tons steel; 3000 tons coal = 1000 tons steel
    d. 60,000 tons coal; 36,000 tons steel
    e. 72; coal, 36; steel; 12

GLOSSARY TO CHAPTER 32

Absolute advantage   The ability to produce more output from given inputs of resources than other producers can.

Comparative advantage   The ability to produce a good or service at a lower opportunity cost than other producers.

Dumping   Selling a good or service abroad at a price below the price charged in the home market or at a price below its cost of production.

General Agreement on Tariffs and Trade (GATT)   An international agreement established in 1947 to further world trade by reducing barriers and tariffs. GATT was replaced by the World Trade Organization in 1995.

Infant industry argument   The contention that tariffs should be imposed to protect from import competition an industry that is trying to get started. Presumably, after the industry becomes technologically efficient, the tariff can be lifted.

Quota system   A government-imposed restriction on the quantity of a specific good that another country is allowed to sell in the United States. In other words, quotas are restrictions on imports. These restrictions are usually applied to one or several specific countries.

Voluntary import expansion (VIE)   An official agreement with another country in which it agrees to import more from the United States.

Voluntary restraint agreement (VRA)   An official agreement with another country that "voluntarily" restricts the quantity of its exports to the United States.

World Trade Organization (WTO)   The successor organization to GATT that handles trade disputes among its member nations.

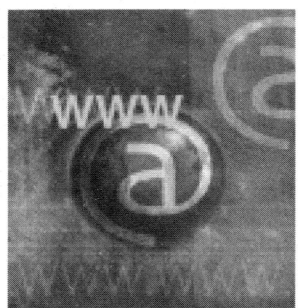

CHAPTER 33
EXCHANGE RATES AND THE BALANCE OF PAYMENTS

LEARNING OBJECTIVES

After you have studied this chapter, you should be able to

1. define foreign exchange rate, accounting identities, foreign exchange market, flexible exchange rates, appreciation, depreciation, gold standard, balance of trade, balance of payments, special drawing rights (SDRs), International Monetary Fund (IMF), dirty float, and par value;

2. distinguish among the balance of trade, the balance on current account, the balance on capital account, and the balance of payments;

3. list three factors that affect a nation's balance of payments;

4. enumerate official reserve account transactions;

5. list transactions that lead to an increase in the supply of a domestic country's currency and an increase in the demand for foreign currency, and distinguish such transactions from transactions that lead to a demand for a domestic country's currency and a supply of a foreign country's currency;

6. recognize the equilibrium exchange rate from graphs, and predict how specific events will change the exchange rate;

7. distinguish between a gold standard international payments system and a flexible exchange rate international payments system.

CHAPTER OUTLINE

1. A balance of trade reflects the difference between the value of a nation's merchandise exports and its merchandise imports; the balance of payments reflects all economic transactions between a nation and the rest of the world.
   a. The current account balance equals the sum of (1) the balance of trade, (2) the balance of services, and (3) net unilateral transfers (private gifts and government grants). If the current

account balance is negative, a current account deficit exists; if it is positive, a current account surplus exists.
- b. Capital account transactions consist of direct investment purchases in financial assets among countries and loans to and from foreigners. If the capital account is a negative number, a capital account deficit exists; if it is positive, a capital account surplus exists.
- c. If the sum of a nation's current account and its capital account is negative, that nation has an international payments disequilibrium (deficit) that must be financed by official (government) reserve account transactions; of course, another nation must have an international payments surplus.
    - i. Official reserve account transactions include sales or purchases of foreign currencies, gold, special drawing rights, the reserve position in the International Monetary Fund, and financial assets held by official government agencies.
    - ii. Official transactions must exactly equal (but be of opposite sign to) the balance of payments.

2. Under a flexible exchange rate system of international payments, exchange rates between nations are determined by the forces of supply and demand.
   - a. When U.S. residents import a foreign good or service, this leads to a supply of dollars and a demand for foreign currency on foreign exchange markets.
   - b. When U.S. residents export goods and services to a foreign country, this leads to a supply of foreign currency and a demand for dollars.
   - c. The equilibrium exchange rate is determined in the same way that the equilibrium price for anything is established.
      - i. The U.S. demand curve for (say) Japanese yen is negatively sloped; as the dollar price of the yen falls—it takes fewer dollars to purchase a given quantity of yen—the quantity demanded for yen, by U.S. residents, rises.
      - ii. The U.S. demand for yen is a derived demand; we demand Japanese yen because we demand Japanese goods and services.
      - iii. The Japanese supply yen because they want (say) U.S. goods and services; as the dollar price of yen rises—it takes fewer yen to purchase one dollar—Japanese residents will increase their quantity supplied of yen in order to purchase more U.S. goods.
      - iv. The equilibrium dollar price per yen is established at the intersection of the U.S. demand for yen curve and the Japanese supply of yen curve; the equilibrium yen price per dollar is automatically determined thereby.
   - d. If U.S. residents experience a change in tastes in favor of Japanese goods, the demand for yen will increase. The dollar price of yen rises and the yen price of dollars falls; the dollar depreciates and the yen appreciates.
   - e. Other determinants of exchange rates include (relative) changes in real interest rates, changes in productivity, changes in tastes, and perceptions of economic stability.

3. Under a gold standard, each nation fixes its exchange rate in terms of gold; therefore all exchange rates are fixed. Under the pure gold standard, each nation would have to abandon an independent monetary policy; each nation's money supply would automatically change whenever a balance of payments disequilibrium occurred.

4. In 1944 representatives of the world's capitalist nations met in Bretton Woods to create a new international payments system to replace the gold standard that had collapsed in the 1930s.
   - a. The International Monetary Fund (IMF) was established in 1944; the IMF established a system of fixed exchange rates and a means to lend foreign exchange to deficit nations.
   - b. Member governments were obligated to intervene in foreign exchange markets, to maintain the values of their currencies within 1 percent of the declared par value.
   - c. In 1971, the United States ended the connection between the value of the dollar and gold, and in 1973 the Bretton Woods system ended.

5. To fix a nation's exchange rate, a central bank must influence the demand for its nation's currency in the foreign exchange market.
   a. To do this, the central bank must buy or sell foreign exchange reserves; if the central bank must persistently sell reserves of foreign currencies to keep the value of its currency at a pegged level, then the exchange rate can remain fixed only as long as the central bank's foreign exchange reserves last.
   b. A key rationale for fixing the exchange rate is to limit foreign exchange risks so that a nation's residents do not have to hedge against losses from fluctuations in exchange rates.
   c. Between a flexible exchange rate system and a fixed exchange rate is a "dirty float," or a managed exchange rate system, in which governments (through their central banks) intervene in foreign exchange markets in order to affect the price of currencies; this is the approach used by most developed nations today.
   d. Some countries have "split the difference" between fixed and floating exchange rates by using a crawling peg, in which the target value of a nation's currency automatically changes over time, or by adopting a target zone, which is an allowed range of permitted exchange rate variations between upper and lower exchange rate bands.

KEY TERMS
Crawling peg
Flexible exchange rates
Foreign exchange market
Foreign exchange rate
Appreciation
Depreciation

Dirty float
International Monetary Fund (IMF)
Gold standard
Balance of trade
Balance of payments
Special drawing rights (SDRs)

KEY CONCEPTS
Accounting identities
Bretton Woods system
Current account

Capital account
Gold standard
Par value

COMPLETION QUESTIONS
Fill in the blank, or circle the correct term.

1. When U.S. residents wish to import Japanese-made goods, they supply _____ to the foreign exchange market and demand _____ on that market; when the Japanese wish to import U.S.-made goods, they supply _____ and demand _____ on the foreign exchange market.

2. In a flexible exchange rate system, exchange rates are determined by (governments, supply and demand); if the exchange rate goes from $0.010 per yen to $0.012 per yen, the dollar has (appreciated, depreciated) and the yen has _____.

3. If the Japanese yen depreciates, it takes (fewer, more) yen to purchase a dollar; this leads to (an increase, a decrease) in the quantity demanded of yen by U.S. residents and (an increase, a decrease) in the quantity supplied of yen by Japanese residents.

4. If U.S. tastes move in favor of European goods, there will be a(n) _____ in the demand for euros; other things being constant, the euro will (appreciate, depreciate) on the foreign exchange market. This eventually will induce U.S. residents to export (less, more) to Europe and import _____ from Europe.

5. Under a pure gold standard, exchange rates (float, are fixed).

6. If the value of U.S. imports exceeds the value of its exports, the U.S. balance of trade will be a (negative, positive) number, and another country's balance of trade must be a(n) _____ number. The United States then is said to have a trade (deficit, surplus), while the other nation has a trade _____.

7. If governments do not intervene, by definition the sum of a nation's balance on current account plus its balance on capital account will equal _____; if governments intervene in the balance of payments process, then the sum of a nation's balance on current account plus its balance on capital account must exactly _____, but be of opposite sign to, its official transactions.

8. Official reserve account transactions involve the following assets of individual countries: _____, _____, _____, _____, and _____.

9. A nation's balance of payments is affected by, among other things, relative changes in that nation's _____ and _____.

10. If the value of a nation's exports is less than the value of its imports, it is running a trade _____; its currency will (depreciate, appreciate) under a flexible exchange rate system.

11. When nations intervene in foreign exchange markets in order to affect exchange rates, a freely floating exchange rate system becomes a(n) _____ float.

12. An automatically adjusting target for a nation's exchange rate is a(n) _____.

13. The possibility that people may incur losses as a result of unexpected variations in the value of a nation's currency in foreign exchange markets is called _____.

TRUE-FALSE QUESTIONS
Circle the **T** if the statement is true, the **F** if it is false. Explain to yourself why a statement is false.

T F 1. If you wish to buy German goods, you ultimately offer dollars and demand euros.

T F 2. If you wish to send money to your relatives in England, you ultimately offer dollars and demand English currency.

T F 3. In a flexible exchange rate system, gold flows lead to international payments equilibrium.

T F 4. If French tastes move in favor of U.S. goods, the supply of dollars on the foreign exchange market rises relative to the demand for dollars.

T F 5. The U.S. demand for British pounds rises if the British inflation rate exceeds the U.S. inflation rate.

T F 6. In a flexible exchange rate system, if Canadian tastes move away from U.S. goods (other things being constant), both the U.S. dollar and the Canadian dollar will depreciate.

T F 7. The gold standard is one form of a fixed exchange rate system.

T F 8. Under the gold standard, if disequilibrium exists in the world's balance of payments, gold will flow from one nation to another until payments equilibrium is restored.

T F 9. Under a flexible exchange rate system, if disequilibrium exists in the world's balance of payments, exchange rates will change until payments equilibrium is restored.

T F 10. Under a flexible exchange rate system, each nation must give up control over its own monetary policy.

T F 11. In today's world, the sum of a nation's current account balance plus its capital account balance must be zero.

T F 12. If one nation has a current account deficit, another nation must have a current account surplus.

T F 13. A nation can finance a current account deficit with a capital account surplus.

T F 14. A dirty float results because nations do not want to pay the price of adjusting to a balance of payments disequilibrium.

T F 15. Under a flexible exchange rate system, payments equilibrium is brought about by a change in the exchange rate; under a gold standard, national price levels change to restore payments equilibrium.

T F 16. To fix, or peg, the exchange rate for its nation's currency, a central bank must buy or sell domestic securities such as bonds issued by the nation's government.

T F 17. Under a target zone approach to managing the exchange rate, a central bank intervenes to keep the exchange rate above an upper band or below a lower band.

MULTIPLE CHOICE QUESTIONS
Circle the letter that corresponds to the best answer.

1. If the foreign exchange rate is that $1 is equivalent to 4 Polish zlotys, then 1 zloty is worth
   a. $4.
   b. 40 cents.
   c. 25 cents.
   d. 4 cents.

2. The demand schedule for yen on the foreign exchange market
   a. is derived partially from foreign demand for Japanese goods.
   b. reflects the fact that Japanese residents want to import goods and services.
   c. shows the quantity of yen demanded at different income levels.
   d. is unimportant if Japan is on a fixed exchange rate system.

3. Which of the following does **NOT** lead to an increase in the demand for Mexican pesos?
   a. A worldwide change in tastes in favor of Mexican goods occurs.
   b. The Mexican inflation rate exceeds the world inflation rate.
   c. Mexico's interest rate rises relative to world rates.
   d. World real income rises.

4. Which of the following leads to an increase in the demand for the U.S. dollar on the foreign exchange market?
   a. an increase in U.S. exports
   b. an increase in foreign investment in the United States
   c. an increase in gifts from foreigners to U.S. residents
   d. All of the above

5. If a nation has an international payments surplus in a flexible exchange rate system, then
   a. its currency will appreciate.
   b. its price level will rise.
   c. gold will flow from it to nations with a payments surplus.
   d. All of the above

6. Which of the following statements is **NOT** true?
   a. Under flexible exchange rates, international payments equilibrium is restored through changes in exchange rates.
   b. Under a gold standard, international payments equilibrium is restored through changes in national price levels.
   c. Under a flexible exchange rate system, a nation cannot pursue a monetary policy that is independent of its trading partners.
   d. Under the gold standard, international payments disequilibrium leads to gold flows, which restore equilibrium.

7. If Thailand has a payments deficit, payments equilibrium can be restored if Thailand's
   a. price level rises relative to the world's.
   b. interest rate rises relative to the world's.
   c. real national income rises relative to the world's.
   d. money supply rises relative to the world's.

8. A nation can finance a deficit on its current account with
   a. a surplus on its capital account.
   b. a deficit on its capital account.
   c. official purchases of foreign currencies with its own currency.
   d. purchases of gold from foreign countries with its own currency.

9. If a nation has a deficit on both its current account and its capital account, then
   a. it is in a balance of payments equilibrium.
   b. the world must be on a flexible exchange rate system.
   c. it must have official transactions that are identical to (but opposite in sign to) the sum of those two deficits.
   d. it will experience gold inflows.

10. A nation's balance of payments is affected by its relative
    a. interest rate.
    b. political stability.
    c. inflation rate.
    d. All of the above

11. The dirty float
    a. has emerged in recent years because nations want less flexible exchange rates.
    b. makes fixed exchange rates more flexible.
    c. is common under a gold standard.
    d. is favored over a pure float by people who want their nation to have a monetary policy independent of its trading partners.

CHAPTER 33: EXCHANGE RATES AND THE BALANCE OF PAYMENTS

MATCHING
Choose the item in column (2) that best matches an item in column (1).

(1)	(2)
a. appreciation	f. rise in one currency's value relative to another's
b. depreciation	g. fall in one currency's value relative to another's
c. fixed exchange rate system	h. balance of payments settlements
d. special drawing right	i. gold standard
e. trade deficit	j. value of imports exceeds value of exports

WORKING WITH GRAPHS

1. Consider a situation in which exchange rates are flexible. Consumers in the United States wish to import a good from Germany, a nation that is part of the European Monetary Union.
    a. Calculate the U.S. price of this good, given the German price of the good and the different exchange rates that might prevail as listed in the table below, and place these calculations in the appropriate column. Calculate the quantity of euros demanded by U.S. consumers in order to purchase the import good at different exchange rates. Enter these numbers in the last column.

Exchange rate ($/euro)	German price of the good	U.S. price of the good	Quantity demanded	Total U.S. euro expenditures
0.80/1	1 euro	_____	90	_____
0.85/1	1 euro	_____	80	_____
0.90/1	1 euro	_____	70	_____
0.95/1	1 euro	_____	60	_____
1.00/1	1 euro	_____	50	_____

By looking at the table above, one can conclude that as it takes more dollars to purchase one euro, the dollar price of the import good will _____.

   b. In the table above, you are given the quantity of the import good at different prices. Graph the demand for euros on the grid provided below.

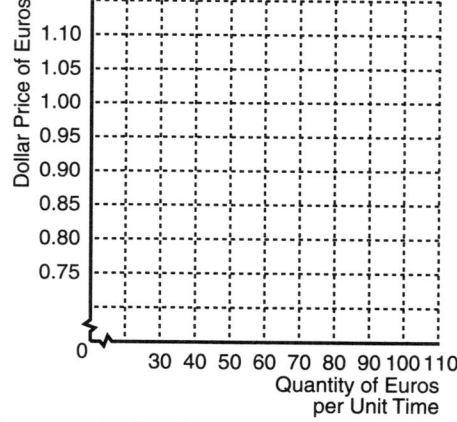

   c. Let us now assume that people in Germany wish to import from the United States some good that costs $1 per unit. Calculate the German price of the good given the U.S. price and the different exchange rates that might prevail, as listed in the table below, and place these numbers in the appropriate column.

Exchange rate ($/euro)	U.S. price of the good	German price of the good	Quantity demanded	Total German $ expenditures
0.80/1	$1	_____	24.0	_____
0.85/1	$1	_____	42.5	_____
0.90/1	$1	_____	63.0	_____
0.95/1	$1	_____	85.5	_____
1.00/1	$1	_____	110.0	_____

By looking at the above table, one can conclude that as it takes more dollars to purchase 1 euro, the euro price of the good will _____.

d. In the last table, you are given the quantity of the import good that German consumers wish to purchase at different German prices. Use this information to calculate the quantity of euros that German consumers will be willing to supply at different exchange rates in order to import the U.S. good. (Note: Round off to the nearest whole number.) Enter these numbers in the last column. Graph the supply of German euros on the same grid as your graph of part b.

e. Assume for simplicity that the only trade between the United States and Germany involves the two goods discussed above. Under this assumption, the equilibrium exchange rate will be approximately _____ dollars per euro or _____ euros per dollar.

f. Suppose that now U.S. consumers undergo a change in tastes and preferences for the German import good. As a result, the U.S. demand for the import good increases as shown in the table below.

Exchange rate ($/euro)	German price of the good	U.S. price of the good	Quantity demanded	Total U.S. euro expenditures
0.80/1	1 euro	_____	120	_____
0.85/1	1 euro	_____	110	_____
0.90/1	1 euro	_____	100	_____
0.95/1	1 euro	_____	90	_____
1.00/1	1 euro	_____	80	_____

Enter in the last column the quantity of euros now demanded by U.S. consumers for use in purchasing the German import good. (Note: Round off to the nearest whole number.) Graph the new demand for euros on the same grid provided for part b.

The new equilibrium exchange rate will be approximately _____ dollars per euro, or _____ euros per dollar. As a result of the increase in the U.S. demand for German imports, with all else constant, the dollar will _____ and the euro will _____.

g. Now consider the above problem assuming that the exchange rate was fixed at $0.90/euro. When the U.S. demand for German goods increased, the United States would have purchased _____ units and paid a total of $_____ for German imports. German residents would have bought _____ units from the United States and paid a total of $_____ for U.S. exports. As a result, the United States would have lost $_____, or approximately _____ euros, in foreign exchange.

2. The figure below shows the supply of, and the demand for, British pounds, as a function of the exchange rate—expressed in U.S. dollars per pound. Assume that Britain and the United States are the only two countries in the world.
   a. How might the shift from D to D' be accounted for?

   b. Given the shift from D to D', what exists at $1.60 = 1 pound?

   c. Will the pound now appreciate or depreciate?

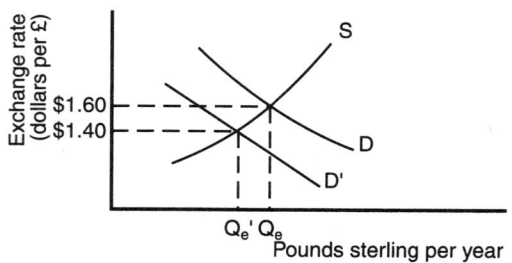

PROBLEMS

1. Below are balance of payments figures for North Shore during 2004. (All figures are in billions of dollars.

Allocations of special drawing rights	$ 710
Balance on the capital account	-3507
Foreign official assets	10475
Errors and omissions	-1879
Balance on the current account	-5795
North Shore official assets	-4

   a. The official reserve transactions balance was (+/-) _____ $_____.

   b. The official reserve transactions were

      (1) the sum of $_____, _____, and _____

      (2) totaled (+/-) _____ $_____

2. Suppose both Canada and the United States had been on a pure gold standard; and the Canadian government had been willing to buy and sell gold at a price of 53.85 Canadian dollars for an ounce of gold, and the U.S. government had been willing to buy and sell gold at a price of $35 an ounce. In the foreign exchange markets, the price of a Canadian dollar would have been _____ U.S. dollars and the price of a U.S. dollar would have been _____ Canadian dollars.

3. Below are hypothetical demand and supply schedules for the European Monetary Union's euro during a week. (The quantities of euros demanded and supplied are measured in millions, and the exchange rate for the euro is measured in dollars.)

Quantity Demanded	Exchange Rate	Quantity Supplied
100	$0.98	570
200	0.97	520
300	0.96	460
400	0.95	400
500	0.94	330
600	0.93	260
700	0.92	180

a. The equilibrium exchange rate for the euro is $_____; and at this equilibrium the rate of exchange for the dollar is _____ euros.

b. At the equilibrium exchange rate,

   (1) _____ million euros are demanded and supplied each week;

   (2) _____ million dollars are bought and sold each week.

c. If the European Central Bank wished to peg the exchange rate for the euro at

   (1) $0.96, it would have to (buy/sell) (how many) _____ million euros for dollars each week;

   (2) $0.94, it would have to (buy/sell) (how many) _____ million euros for dollars each week.

4. Which of the following will cause the yen to appreciate?

   a. U.S. real incomes increase relative to Japanese real incomes.

   b. It is expected that in the future the yen will depreciate relative to the dollar.

   c. The U.S. inflation rate rises relative to the Japanese inflation rate.

   d. The after-tax, risk-adjusted real interest rate in the United States rises relative to that in Japan.

   e. U.S. tastes change in favor of Japanese-made goods.

ANSWERS TO CHAPTER 33

COMPLETION QUESTIONS

1. dollars; yen; yen; dollars
2. supply and demand; depreciated; appreciated
3. more; an increase; a decrease
4. increase; appreciate; more; less
5. are fixed
6. negative; positive; deficit; surplus
7. zero; equal
8. foreign currencies; gold; SDRs; reserve position in the IMF; any financial asset held by an official government agency
9. inflation rate; political stability
10. deficit; depreciate;
11. dirty
12. crawling peg
13. foreign exchange risk

TRUE-FALSE QUESTIONS

1. T
2. T
3. F Changes in exchange rates lead to payments equilibrium.
4. F The demand for dollars rises relative to the supply of dollars, because the French want to buy relatively more U.S. goods.
5. F U.S. residents will demand fewer pounds because British goods are now relatively higher priced.
6. F The U.S. dollar will depreciate relative to the Canadian dollar; the Canadian dollar therefore must appreciate relative to the U.S. dollar.
7. T
8. T
9. T
10. F Floating exchange rate systems permit an independent monetary policy.
11. F In today's world, nations intervene in exchange markets; hence international settlements among governments are necessary.
12. T
13. T
14. T
15. T
16. F The central bank must buy or sell foreign exchange reserves.
17. F Interventions would keep the exchange rate between the upper and lower band, or within the target zone.

MULTIPLE CHOICE QUESTIONS

1.c; 2.a; 3.b; 4.d; 5.a; 6.c; 7.b; 8.a; 9.c; 10.d; 11.a.

MATCHING

a and f; b and g; c and i; d and h; e and j

WORKING WITH GRAPHS

1. a. U.S. prices of the good: 0.80, 0.85, 0.90, 0.95, 1.00. Total U.S. euro expenditures: 90, 80, 70, 60, 50; rise
    b. See graph below.
    c. German prices of the good: 1.250, 1.176, 1.111, 1.053, 1.00; Total German dollar expenditures: 24.00, 42.50, 63.00, 85.50, 110.00; fall
    d. Total German expenditures in euros: 30, 50, 70, 90, 110. See graph below.
    e. 0.90, 1.11
    f. Total U.S. euro expenditures: 120, 110, 100, 90, 80. See graph below. 0.95; 1.53; depreciate; appreciate
    g. 100; 111.10; 63, 63.00; 48.10 (or $111.10 - $63.00), 53 (48.1 ÷ $0.90 = 53.44).

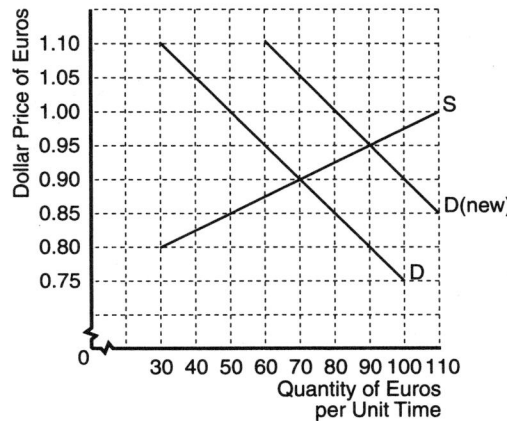

2. a. A decrease in the demand for British pounds will occur if world tastes change away from British-made goods, the British price level rises relative to the world's, world income falls relative to British income, and/or British interest rates fall relative to world interest rates, among other reasons.
    b. A surplus of British pounds, and a British balance of payments deficit.
    c. depreciate

PROBLEMS

1. a. -, 11181;   b. (1) 710, 10475, -4 (any order)   (2) +, 11181
2. 0.65, 1.538
3. a. 0.95, 1.053;   b. (1) 400 (2) 421.05;   c. (1) buy, 160 (2) sell, 170
4. a, c, and e

GLOSSARY TO CHAPTER 33

Accounting identities Values that are equivalent by definition.

Appreciation An increase in the exchange value of one nation's currency in terms of the currency of another nation.

Balance of payments A system of accounts that measures transactions of goods, services, income, and financial assets between domestic households, businesses, and governments and residents of the rest of the world during a specific time period.

Balance of trade The difference between exports and imports of goods.

Capital account A category of balance of payments transactions that measures flows of real and financial assets.

Crawling peg An exchange rate arrangement in which a country pegs the value of its currency to the exchange value of another nation's currency but allows the par value to change at regular intervals.

Current account A category of balance of payments transactions that measures the exchange of merchandise, the exchange of services, and unilateral transfers.

Depreciation A decrease in the exchange value of one nation's currency in terms of the currency of another nation

Dirty float Active management of a floating exchange rate on the part of a country's government, often in cooperation with other nations.

Exchange rate The price of one nation's currency in terms of the currency of another country.

Flexible exchange rates Exchange rates that are allowed to fluctuate in the open market in response to changes in supply and demand. Sometimes called *floating exchange rates*.

Foreign exchange market A market in which households, firms, and governments buy and sell national currencies.

Foreign exchange risk The possibility that changes in the value of a nation's currency will result in variations in the market value of assets.

Hedge A financial strategy that reduces the chance of suffering losses arising from foreign exchange risks.

International Monetary Fund A multinational organization that aims to promote world economic growth through more financial stability.

Par value The officially determined value of a currency.

Special drawing rights (SDRs) Reserve assets created by the International Monetary Fund for countries to use in settling international payment obligations.

Target zone A range of permitted exchange rate variations between upper and lower exchange rate bands that a central bank defends by selling or buying foreign exchange reserves.